represents that nation's past. Galuszka's *Thunder on the Mountain* highlights the disturbing and often deadly impacts of this highly polluting energy source and why Big Coal might just be losing its power."

—Ron Pernick, managing director of Clean Edge, Inc., and coauthor of *The Clean Tech Revolution* and *Clean Tech Nation*

"With measured and dogged reporting, Galuszka persuasively reveals how corporate greed and mismanagement, Appalachian underdevelopment, insatiable global demand for coal, and the right-wing backlash against government regulation and labor organization resulted in tragedy at Upper Big Branch. Essential reading for anyone interested in the past, present, and future of Big Coal."

—Thomas G. Andrews, Bancroft Prize–winning author of *Killing for Coal: America's Deadliest Labor War*

"*Thunder on the Mountain* is an important book about a coal-mining disaster, and it is also a timely reminder of the dangers of putting profits before safety in the energy business."

—Stanley Reed, former London bureau chief of *BusinessWeek* and coauthor of *In Too Deep: BP and the Drilling Race That Took It Down*

"Peter Galuszka exposes a seam that runs deep in American history—the corporate indifference of Big Coal, its neglect of worker safety, and the fight waged by miners and their families for dignity and quality of life."

—Philip Dray, author of *There Is Power in a Union: The Epic Story of Labor in America*

"Peter Galuszka has a deep appreciation of Appalachia and its damaged beauty, having grown up in West Virginia and filed decades of coal stories as a reporter. In *Thunder on the Mountain* he draws vivid portraits of all the characters in the ongoing tragedy of Appalachian mining, from the twenty-nine victims of Upper Big Branch to the watery-eyed, self-righteous CEO of Massey Mining, Don Blankenship, whose brutal ways brought such misery to so many."

—Michael Shnayerson, author of *Coal River*

"Peter Galuszka has written a powerful book that lays bare the corporate greed behind one of the worst mining disasters in modern times. *Thunder on the Mountain* puts a human face on tragedy, and Galuszka's own ties to West Virginia provide poignant context to both mine workers' plight and the environment for which Massey Energy showed what can only be described as contempt. Every member of Congress should read this book and then ask themselves why they failed to pass a mine-safety bill in the wake of such unabashed disregard for safety and human life."

—Loren C. Steffy, author of *Drowning in Oil: BP and the Reckless Pursuit of Profit*

"Appalachia may be blessed with the 'world's best metallurgical coal,' but as journalist Galuszka's powerful book shows, this coal is both a 'curse and a prize'. . . . He convincingly excoriates the safety record of Massey Energy and its controversial former CEO, Don Blankenship. . . . Drawing on his personal experience of Appalachia, Galuszka offers a sympathetic but unsentimental portrait of the region's people and their struggles."   —*Publishers Weekly*

THUNDER ON THE
MOUNTAIN

Death at Massey
and the Dirty Secrets
Behind Big Coal

PETER A. GALUSZKA

St. Martin's Press ≈ New York

www.stmartins.com

All photographs except the following are courtesy of Scott Elmquist: Insert, page 2, bottom: Scott J. Ferrell, Getty Images. Insert, page 4, top: Chip Ellis, *The Charleston Gazette*.

Design by Anna Gorovoy

Library of Congress Cataloging-in-Publication Data

Galuszka, Peter A.
    Thunder on the mountain : death at Massey and the dirty secrets behind big coal / Peter A. Galuszka.
        p. cm.
    Includes index.
    ISBN 978-1-250-00021-7 (hardcover)
    ISBN 978-1-250-01808-3 (e-book)
    1. Coal trade—Appalachian Region.   2. Coal mines and mining—Appalachian Region.   3. Massey Energy (Firm)   I. Title.
    HD9547.A67G35 2012
    363.11'96223340974—dc23

                                                                    2012028241

First Edition: September 2012

10   9   8   7   6   5   4   3   2   1

For Marina and our children, Camille and Maria

CONTENTS

CONTENTS

ACKNOWLEDGMENTS

My first book effort benefited tremendously from the help of many people. My thanks to the families of the victims of the Upper Big Branch disaster, who endured painful moments to help me understand what had happened and how it had reshaped their lives.

In West Virginia, special thanks go to Patty and Wayne Quarles, Tommy Davis and his family, and Terry Ellison. They were always hospitable and generous although the issues we discussed were emotionally hurtful. Vernon Halmton met with me on several occasions over many months to discuss mountaintop removal. Country music artist Kathy Mattea took time for an interview, as did members of the Taylor Made country rock group. Although I have never met him, Ken Ward of the *Charleston Gazette* provided consistently high-quality reporting of Massey Energy. National Public Radio

ACKNOWLEDGMENTS

had intelligent analysis that went beyond the daily news. Gary Stover provided technical insights.

Closer to home, in Richmond, E. Morgan Massey spent hours with me helping me understand the roots of Massey Energy while knowing that my book would be less than glowing. Former Massey directors Stanley C. Suboleski and James Crawford provided insightful interviews. My good friends Jim Bacon and Frank Green helped me make contacts and kept me in the news loop. *Style Weekly*, an alternative publication, published my story in 2009 about mountaintop removal that was a precursor for this book. Thanks go to Scott Bass, Jeff Bland, Jason Roop, and Lori Waran, and special kudos to senior contributing editor Edwin Slipek Jr. for helping me understand Richmond's social history.

In the Virginia coalfields, Kevin Crutchfield, head of Alpha Natural Resources, and his staff were generous with interview time and let us visit one of his mines. In Norfolk, Robin Chapman of Norfolk Southern Railroad graciously gave our photographer and me access to their coal-export facilities. In Austin, Texas, Admiral Bobby Ray Inman filled me in on the boardroom view.

In Asia, my wife, Marina, then teaching in Shanghai, helped set up travel and accompanied me on the Chinese and Mongolian legs of my research journey. In Tokyo, *Business-Week* friend Toshio Aritake was most generous. In the Washington area, Casey Aiken, one of my oldest friends, offered savvy legal advice.

In New York, another former *BusinessWeek* colleague, Bill Holstein, was encouraging from the very first moment. One couldn't ask for a better editor than Matt Martz at St. Martin's Press, who patiently guided me past my organizational and verbosity excesses. Scott Elmquist, a colleague from *Style* and a brilliant photographer, cheerfully put up with miles on the road, a trip in a deep mine, and a choppy airplane ride to get the best coalfield imagery you will ever see.

Lastly, many thanks to my outstanding and tireless agent, Leah Nathans Spiro, president of Riverside Creative Management. Another former *BusinessWeek* colleague, Leah backed this project from the start, served as an excellent sounding board, and provided indispensable guidance.

PREFACE

Appalachian coal has been a part of my life for years, however tangentially. My introduction occurred in 1962, when I was nine years old. My father, an altruistic navy doctor in Bethesda, Maryland, decided to retire from his military career and took us to central West Virginia, a place where we had no family or other ties. He had other opportunities to practice medicine in more sophisticated and better-paying locales, but he wanted to work where he could do the most good. My sister and I were shipped from one of the most affluent counties in the country to one of the poorest. My fifth-grade arithmetic book had been in use since 1903. I rode school buses with coal miners' children over endlessly twisting mountain roads. We had been modestly comfortable on Dad's navy pay in the D.C. area. In Harrison County, West Virginia, where the "War on Poverty" was declared by Lyndon B. Johnson in 1964, we were considered rich.

PREFACE

I used to play on strip mines near my home, skirting the skeletons of small animals next to rain ponds made toxic by exposed coal because of lax regulation. On occasion, a schoolmate's dad would die when a heavy piece of a coal mine's roof, usually a slab of slate, would fall. We usually would learn of a death only after a child didn't show up on our school bus.

Years later, my journalism career took me back to the area. One visit was in 1979, when energy prices spiked so high that one hundred coal ships swung at anchor, waiting for coal dock space at Hampton Roads, Virginia, where I worked for a newspaper. My first job with McGraw-Hill in 1983 was working for a Washington-based coal-industry newsletter that had me hopscotch the United States visiting coal mines, including ones in Central Appalachia. In 2002, I interviewed Don Blankenship, a subject of this book, who graciously let me visit a Massey Energy deep mine that was "low coal," meaning its seams were about forty-nine inches tall, forcing miners to work in a stooped-over position for hours.

I have come to know coal's dangers firsthand and the culture it comes from. I care about the people who make their living from it and the human lives, mountains, and streams it destroys. I wanted to write this book to help me understand what it was about the Appalachians that so intrigued me. The contradictions are enormous: so much potential wealth but, in reality, so much poverty; beautiful, rugged landscapes marred by miles of tan strip-mine gashes; and people and ideas isolated by geography and culture. I want to explain it to others.

THUNDER ON THE
MOUNTAIN

INTRODUCTION

For Appalachia, coal is a curse and a prize. Since the nineteenth century, it has built cities, electrified houses, and been used for transportation and synthetic energy. It has also killed thousands of miners, ruined many lungs, polluted many streams, and bitten off huge chunks of some of the most beautiful mountains in the United States. Although coal's uses have shifted over the years—railroads no longer run on it nor are homes heated with it—more coal is produced than ever before, providing nearly half of all electricity generated in twenty-first-century America and building skyscrapers and highways in Shanghai and Mumbai.

Coal's contradictions come together most tellingly in the form of one man, Donald Blankenship, a child of the Central Appalachians who was chairman and chief executive officer of America's fourth-largest coal firm, Massey Energy. Like the rawboned settlers in the hills of southern West Virginia and Eastern Kentucky, Blankenship was tough, independent minded, and not at all beholden to whatever latest management fad was coming out of business schools. Even now, Blankenship makes no bones about breaking unions, telling environmentalists where to get off, bankrolling mountain politicians and judges, and resisting complaints of do-good "green investors" who were critical of Massey's poor safety record and devastating mountaintop-removal practices.

Although Blankenship was forced to retire and his company was taken over by a competitor after a massive explosion killed twenty-nine miners at Massey's Upper Big Branch Mine in 2010, the tensions of the industry remain. Coal remains exceptionally well positioned to continue as a major source of electrical power, given the Fukushima nuclear power plant disaster in Japan and the dearth of carbon-control legislation. Another dynamic is also driving coal—the insatiable demand by the fast-growing Asian countries such as China and India, which need great amounts of metallurgical (or met) coal to make coke and steel.

For the coal industry, its list of dirty secrets prevails. The industry doesn't want people to know how it is destroying mountains and fragile ecosystems with its mountaintop-removal practices. Its propaganda says that it is taking necessary safety and environmental risks to keep America's electric lights on, when, in fact, much of the coal mined in the region, and notably at Upper Big Branch, has absolutely nothing to do with electricity generation. It is bound for the international market to make steel in foreign countries. Despite the abuses that led to the twenty-nine dead at Upper Big Branch, the coal industry and its well-financed lobbyists have effectively killed any meaningful legal and regulatory reforms that would take such needed measures as having boards of directors and top corporate officials of coal firms held criminally liable if they know of unsafe mining conditions and take no steps to correct them. As the usual way of doing business prevails, the cycle of injustice endures. The world's best metallurgical coal comes from Central Appalachia, ensuring big profits for coal operators, yet the region remains among the poorest in the country, as it always has been.

1

DEATH AT UPPER BIG BRANCH

In the morning darkness of April 5, 2010, Tommy Davis left his home with its cluttered yard for work. A sinewy, forty-three-year-old who rides Harleys and hunts black bears with a bow and arrow, Davis had worked at a Massey Energy Company surface mine for twelve years. Four months before, drawn by higher pay and the chance to work with as many as five of his relatives, including Cory, his son, he had transferred to Massey's Upper Big Branch deep mine about forty-five minutes away in Raleigh County in the Coal River Valley.

The day before had been Easter Sunday. Work at Upper Big Branch had been suspended so miners could enjoy their usual paschal activities. Families attended church, searched for Easter eggs with their children, and ate baked ham dinners. An early shift had started at midnight but did little more than maintenance work. The first regular production run began at 6 A.M. Dawn that Monday, April 5, promised temperatures in

5

the 70s, unusually warm for the fickle early spring of southern West Virginia. There, snow showers and wind quickly change back and forth into sunny days that bring out ramps, a wild-growing onion with a pungent garlic aroma that is a seasonal delicacy in this part of Appalachia, and the chirping of spring peeper frogs.

Known as UBB, the fifteen-year-old mine is nestled on the west side of a narrow valley marked by Coal River, which after spring rains is a brown, wildly churning stream capped by small wavelets of white water. Potato chip bags, bits of clothing, and other trash cling to tree limbs after floods push the river over its banks. Next to it, on a CSX Transportation rail branch line, hopper cars clatter in for loading at the valley's half a dozen or so mining operations. Scattered here and there amid the hardwood trees and rock outcrops are reminders of just how hazardous coal work can be. Occasional roads of industrial-grade gravel leading to coal mines have signs boldly lettered AMBULANCE ENTRANCE.

UBB was operated by Performance Coal, one of Massey Energy's more than forty subsidiaries that had been over-seen by former Massey Energy board chairman and chief executive officer Donald Blankenship, who is a tall, jowly man with a prominent double chin and steady, penetrating stare. Coal River, about thirty miles south of the state capital of Charleston, has been the epicenter of a drama that has featured the highly controversial Blankenship for more than a decade. A native of Central Appalachia, Blankenship parlayed an extraordinary gift for crunching numbers with an indefatigable work ethic he learned from his single mother into becoming Big Coal's best-known, and most notorious, corporate executive.

As he did at many of his company's operations, Blankenship flew around in a company helicopter like a twenty-first-century William Westmoreland, the celebrity general of Vietnam War fame whose penchant for statistics and body counts has become legend. Blankenship moved about, checking production numbers here and solving minor problems there, such as micromanaging when and if an overtime shift got a lunch break. He spent much of his time pushing faster, more efficient, and cheaper production, demanding reports of the output of each mine several times during each work shift. He battled safety and environmental regulators; bankrolled state political candidates who favored the coal industry, including judges; and waged an intense public relations war against ecological activists, whom he despised and dubbed "greeniacs." His cantankerous ploys were often successful. He was effective in his move to block national legislation stemming coal-related emissions of greenhouse gases that contribute to climate change. His tenacity prevailed when a West Virginia judge ruled that big concrete silos filled with coal did not harm mountain schoolchildren at the Marsh Fork Elementary School near Massey's huge Edwight "mountaintop removal" surface mine in the Coal River Valley. This same mine also has a 3.8-billion-gallon pond of dark toxic sludge from mine tailings held back by an earthen dam high above the school. While Massey eventually contributed to build a new school at a different location, children for years endured the threat of breathing carcinogenic compounds from coal dust and drowning from a possible dam break.

In this type of mining, which became widespread in southern West Virginia and Eastern Kentucky over the past two decades and has been targeted by celebrity protests and local ecologists who regularly employ guerrilla tactics to snare media attention, hundreds of feet of dirt, rock, and trees— "overburden" in mining company parlance—are lopped off like the cap of a Coca-Cola bottle by powerful explosives and gigantic drag lines.

UBB is a deep mine situated a few miles north of Edwight on the same side of the road. Unlike strip mines on the surface, deep mines can run thousands of feet into the earth. The aboveground section of the deep mine is a tangle of metal buildings, conveyor belts, and tall supporting towers that jut up dramatically from the narrow valley. The mine is of crucial importance to Massey because it taps the Eagle Seam of incredibly rich metallurgical coal that is in tremendous demand, especially in Asian countries such as China that are in a construction boom, building skyscrapers, bridges, and high-speed passenger trains. Despite the Great Recession, demand had started spiking for metallurgical coal in late 2009 and kept pace through the following two years. During the first half of 2010, met coal exports from the United States would reach

39.8 million tons, a 62 percent increase. Another advantage of Eagle Seam coal, as with most of the product in southern West Virginia and Eastern Kentucky, is that it can also be used to fire giant steam generators at electric power plants in the United States, which depend on coal for at least 45 percent of their electricity.

Despite its long-standing reputation for cutting corners and costs, Massey Energy was struggling to catch up with the unexpected Asian boom. In 2009, Upper Big Branch was cited by the federal Mine Safety and Health Administration 515 times for safety violations, nearly twice the national average. It was fined a total of $382,000 just for UBB in 2009. During the previous month alone, UBB had been closed for safety violations more than sixty-one times by MSHA—more than any other mine in the country. Massey officials had a perpetual feud with MSHA over changes in ventilation plans at the mine. Since January 2009, the mine had been cited forty-eight times for air-related problems. Miners were fearful of the erratic airflows and the unusually high number of air-lock doors. These are like watertight doors on a ship and can be shut to manipulate airflow. They are cheaper to install than other, safer types that don't break the airflow pattern when opened and are less prone to being opened or closed unintentionally. At UBB, the doors were constantly being opened for work. Air is critical in any mine, but it was especially important at a huge one like Upper Big Branch, whose geology was unusually "gassy" with methane and whose shafts stretched for miles underground. Workers in such mines are more prone to injury or death by being ripped apart or crushed by debris from an explosion, burned by fire, or suffocated by toxic gases such as methane or carbon monoxide.

Many claim that Massey Energy was under intense financial pressure to produce, since its stock price had been sagging. According to a lawsuit filed on April 29, 2010, by the Macomb County Employees Retirement System, an institutional investor in Massey stock, "the number of violations at Massey mines had dramatically spiked in 2009 as the Company ramped up production attempting to reverse a year-long slide in profitability during which its stock price had collapsed from more than $80 per share to as low as $10 a share."

Still, to miners like Tommy Davis, Massey was a godsend. Unemployment was running better than 10 percent in Raleigh and surrounding counties. If there were jobs in the tiny burgs that dot the hollows, they tended to be at gas stations, pizza joints, or the ubiquitous Dollar General stores offering cheap merchandise. Fast-food jobs like McDonald's can be thirty miles away, and low-paying work at a Walmart farther still in places such as Charleston or Beckley. Mining, by contrast, paid upward of sixty-eight thousand dollars a year, or more than double the state average annual salary. Deep-mining could pay even more. "You might make twenty-four dollars an hour at the surface mine, but in a deep mine, I make thirty-one an hour. That's a hundred forty to a hundred fifty dollars a day more," Tommy Davis said. The extra money was a big help when it came to paying the bills and raising children, not to mention his love of motorcycles and pickup trucks, which sit in his yard.

On that Monday after Easter, Tommy Davis parked his car at the mine and hopped aboard a mantrip, a kind of low-slung truck or railcar that can carry up to thirteen miners stuffed aboard with their helmets; battery packs; metatarsal-protective, steel-toed boots; and self-rescuers—temporary breathing apparatuses used if the mine becomes smoky or otherwise short of air. The shift began with problems. After being closed for the holiday, some sections of the mine had been flooded by underground water. Workers dealing with pumps had gone to work wearing long johns and heavy pants, since they expected to work in the cold. "You would have a thermal shirt on, a jacket, gloves, or a beanie . . . but that day was miserably hot," miner David Farley recalls. Some men even stripped to their shorts. Another oddity: air seemed to be flowing in an opposite direction that morning. Miners later recalled it being a telltale sign that something wasn't right.

One miner who seemed especially spooked going to work that day was Gary Wayne Quarles, a thirty-three-year-old man so large and round at three hundred pounds that he was nicknamed Spanky. He had just gone through a painful divorce and was staying with his father, also a miner and also named Gary, and his mother, Patty, at their trim double-wide trailer home off a small creek near Naoma not far from Upper Big Branch. Besides doting on his eleven-year-old son and nine-year-old daughter, Quarles enjoyed hunting for deer and wild turkey with his father and friends. The night before, he had gone out to a Hooters restaurant in Beckley. With him were Jason Gautier, a former Massey employee then working with another coal firm, and Nicolas McCroskey, also a Massey worker. Quarles was morose at the meal. He and McCroskey told their colleague that "something bad was going to happen" at Upper Big Branch. The next morning, the younger Quarles anxiously went to his job.

The same morning, Tommy Davis's mantrip entered the mine at a downward angle and traveled miles into and hundreds of feet beneath the mountain surface. Going to work was a family affair for Davis, and then some. Of the sixty-one miners working the 6 A.M. to 3 P.M. shift, there was his twenty-one-year-old son, Cory, and his brother Timmy. A nephew, Josh Napper, was a newcomer who had moved in with his grandparents not far from Dawes to work the coalfields because he had lost his job as a registered nurse in south central Ohio. Another nephew, Cody, was also on the shift.

The mantrip ride took about half an hour and transported the miners nearly five miles into the mountain to a longwall mining apparatus. Other miners went to another section hundreds of yards away that was being prepared for a repositioning of the longwall device at a later date. Considered the most efficient and profitable method of deep-mining, the longwall is a massive and expensive drilling rig that runs one thousand feet, back and forth, ripping out coal. Its spearhead consists of two devices called "shearers," which are covered with ultra-hard bits and 158 water-spray nozzles to keep coal dust down. The shearers roar back and forth, up and down, a seam. At one end, called a head end, coal is pushed onto conveyors, belts that whisk it miles to the surface, where it can be classified, washed, and prepared for shipment by railcar or truck to domestic or global customers. When the device reaches the "tail end," it reverses course and moves back to the head end again, screaming and straining as it rips out big chunks of black coal. Typically, the mined area, held up by hydraulic jacks from the tremendous force of the mountain bearing down on it, is eventually buried as the jacks are moved and the mine roof collapses behind it after miners and machinery are moved away. The longwall device then moves ahead, eating into the mountain. Gary Wayne Quarles was one of several miners operating the device that day.

Davis said it seemed a routine shift. He spent part of his time laying track in the area where the longwall was due to be placed. "It's pretty low, maybe fifty inches in the highest part, and I'm six foot four inches tall," he said. "In some sections, you have to crawl on your hands and knees." Some of his relatives were on roof-bolting assignment that involved pounding metal bolts the size of car hubcaps into the mine ceiling to hold up the roof. According to a MSHA report, Massey supervisors on the surface got a call from miners near the coalface at 7:30 A.M. About 11 A.M., the longwall machine ran into a problem and shut down. A retainer holding a hinge for a ranging arm had come loose. Without it, the longwall could not operate. That cost money for Massey, then struggling to boost its stock price after production flaws had tanked it to the ten-dollars-a-share level. After repairs and tests, the longwall machine resumed operation at 2:15 P.M., toward the end of the shift. Up top, miners started to prepare for the next shift as they gathered their gear, including their heavy mine jackets marked by fluorescent orange and silver stripes—Massey Energy's colors.

Down below, around 2:30 P.M., Davis and a nephew quit their shifts a little early and started heading for the surface on the mantrip, stopping for a few moments to chat with his son and others. It was, he said, the usual macho miner camaraderie "trying to get each other's goat." He and his nephew were about two hundred feet from the surface when he suddenly felt the wind at his back. His nephew jumped and took shelter in front of the mantrip, and Davis started running for the opening. Moments later came a second whoosh of air, this one far more powerful. "I felt this wind and all this shit coming out—rocks and wood. I made it outside and was trying to get my bearings. I thought it was a major rockfall, but then I remember them all back there: my son, my brother, my nephew, and the others."

Another surviving miner, Steven Smith, described the experience this way: "Before you knew it, it was just like your ears stopped up, you couldn't hear, and the next thing you know, it's just like you're in the middle of a tornado."

A massive explosion had ripped through UBB's maze of shafts, headgates, tailgates, and mining rooms, rolling at least seven miles underground, turning abruptly at right angles along shafts and, at times, looping around and inundating the same spaces twice. MSHA investigators believe the blast started when the longwall machine hit a stretch of sandstone. Shearer bits on the longwall machine created a big splash of sparks as it hit. "Coal and shale are soft and the shearer can bore right through them. But sandstone sparks quite a bit," says Gary Stover, a former Massey mine engineer who now brokers coal-land deals for Penn Virginia Resource Partners in Chesapeake, West Virginia. The sandstone was apparently so tough that some of the carbide-tipped bits on the shearer had been stripped down to bare steel, further increasing the chances of spitting out sparks.

MSHA officials believe that at 3:02 P.M. a small bubble of methane gas roughly the size of a basketball shot out from the coalface as the shearer hit sandstone. Sparks from the shearer ignited it. The flame blossomed for up to ninety seconds. Incredibly, someone had shut off the water sprayed by the jets in the shearer that are designed to quell flames in exactly this type of situation. During that critical minute and a half or so, the methane flames touched off loose coal dust that was found at high levels through the various mine shafts. In the right mix of air, coal dust can be as explosive as trinitrotoluene (TNT). Water sprays, robust air ventilation, and limestone coatings are used to keep it in check. None worked.

Terrified miners scrambled for their lives, crawling away as fast as they could. One of them was Gary Wayne Quarles, who had been operating the longwall with Joel Price, Christopher Bell, and Dillard Persinger. They obviously knew something was terribly wrong, because they tried to get away quickly. But it was no use. The tremendous blast forces roared to the south away from them but made two left-hand, 90-degree turns at shafts and then zipped back within seconds to engulf them. Their bodies were found about a third of the way down the longwall headgate. The blast forces fed on themselves and raced through the underground corridors with such force that equipment was smashed and some miners were decapitated. Tommy Davis's son, brother, and nephew were killed in the new headgate section, thousands of feet from the source of the explosion. In another section, the remains of Nicolas McCroskey, the miner who had had supper with Gary Wayne Quarles at Hooters the night before the blast, were smashed onto a shaft ceiling. The body wouldn't be found for several days, even though rescuers had passed by several times. McCroskey was discovered only when the stench of his rotting body forced rescuers to look up and not down as they had been doing.

The blast ripped through the longwall room and into various shafts around it. What miners weren't killed by the blast trauma suffocated when the powerful explosion sucked air out of the mine shafts and replaced it with toxic carbon monoxide. It happened so quickly that miners didn't have the mere seconds of time it took to don self-rescuer face masks.

Later, when poisonous gases had cleared enough to allow investigators to enter, they found that the shearer bits were worn and water sprays on it were not working. Among its many accusations against Massey, MSHA claimed that the firm had failed to spread enough noncombustible crushed limestone on surfaces in the shafts to prevent just such a coal-dust explosion. Just the month before, MSHA had cited Massey for not properly ventilating methane gas at the mine, and coal dust had been a problem at other Massey operations. Massey officials dispute MSHA's claims, saying that photographs taken at the coalface later revealed that curious slits had opened up at the mine-room floor near the tailgate section. This suggests that somehow cracks in the earth had split open—perhaps from a natural seismic event—and methane gas from a coal seam beneath the one being mined wafted up and exploded. If so, it was an event out of Massey's control. As late as June 2010, two months after the blast, Bobby Ray Inman, a retired navy admiral who is the former head of the ultrasecret National

Security Agency and was lead independent director at the time of the blast and chairman of the Massey board of directors after Blankenship was forced out, was still claiming that the UBB blast was "an act of God."

Up on the surface, the blast sparked mass confusion. Several miles to the south of UBB on Route 3, the only road to the mine, Gary Jarrell was working the cash register at the 125-year-old Jarrell General Store. It's a classic country outlet selling cans of pork and beans, bread, soft drinks, and cigarettes. Managers cash paychecks for trusted customers. On that afternoon, Jarrell says he didn't hear anything and doesn't remember the time, but he suddenly noticed that one, then two ambulances were racing northward on the highway. "That was not unusual, we thought there might have been a wreck. But they kept coming and coming. Then police cars. Then fire engines. Rumors were flying about which mine it was."

As they had under Blankenship's leadership for years, Massey officials instinctively circled their wagons. According to *Charleston Gazette,* at 3:30 P.M., for example, an unidentified Massey official called the state's industrial accident hotline to report "an air reversal on the beltlines" and increased levels of carbon monoxide. The mine was being evacuated. "Thank you, sir," the hotline operator said. "You have a nice day."

"You do the same," replied the Massey official.

There were conflicting reports of what had happened and

when. MSHA said the blast was touched off at 3:02 P.M., but Massey's internal monitoring data shows that carbon monoxide alarms started going dead six minutes later. Massey officials claim that they had a rescue team at the mine by 4 P.M., but state and federal regulators said that Massey didn't contact them about the blast until 3:27 P.M, so it would take at least another hour and a half before federal rescue teams were there. Consol Energy, a major mining firm, immediately sent its own company rescue team, but the crew left in disgust, saying that with the sloppy way Massey Energy officials were handling the rescue, it would be too dangerous to stay. Lynn Seay, a spokeswoman for Consol, says, "based on information our internal experts were able to ascertain throughout the week of April 5, 2010, at UBB, Consol Energy made the call not to participate in the operations."

Chris Blanchard, president of Performance Coal; Jason Whitehead; and two other Massey managers were in the mine for several hours. None had appropriate mine rescue training. Questions were raised regarding what their purpose was in the mine when their presence violated rescue safety protocols. When questioned later by regulators why they had entered the mine and what they did while there, both Blanchard and Whitehead invoked their Fifth Amendment rights against self-incrimination.

One of the survivors they came across was Timothy Blake, a roof bolter with thirty-eight years' experience who had spent a year at Upper Big Branch. He had been on a mantrip leaving his shift with Steve Harrah, James Woods, Bill Lynch, Carl Acord, Jason Atkins, Benny Willingham, Robert Clark, and Deward Scott. Speaking of the explosion, he said, "Everything just went black. It was like sitting in the middle of a hurricane, things flying, hitting you." He says he struggled to get his self-rescuer and air unit on, but in the mayhem and toxic air, he couldn't find his goggles. He heard gurgling. "It was my buddy beside me, [Jason Atkins,] the twenty-three-year-old boy. . . . He couldn't get his rescuer on." Neither could others on the mantrip. Blake frantically tried to help. He stayed with the others, waiting for nearly an hour, but his air supply was becoming dangerously low. He had to move or die. He felt for pulses on the miners around him. All had a pulse except for one. "I had to leave. It was the hardest thing I ever done."

Staggering up the mine shaft toward the surface, Blake ran into the Blanchard group of Massey executives, who, without waiting for a proper rescue team, raced into the mine. They had been working their way through the mine but were stopped when they had to remove debris. "We didn't know what we had, so we was just trying to be careful and watch exactly everything as we went," says Pat Hilbert, a foreman with the Blanchard crew. Then he says, "they saw a single light walking towards us." It was Blake, dazed and stumbling. Hilbert found the group Blake had left on a mantrip. By then, six were dead and one later died after being removed from the mine. The rescuers were trying to reach a rescue chamber that could be sealed airtight in an emergency near the longwall operation. When they did, they found more dead miners. By one account, some bodies were so mutilated that the first team walked past them without recognizing them.

The firm issued a press release about the explosion just before 5 P.M., saying that an event had happened and information about miners was "uncertain." About a quarter hour later, a dispatcher with the state Homeland Security group called Jeff Gillenwater, Massey's media spokesman, begging for details. "My director is all over my backside wanting information," he told Gillenwater. Known for blunting media inquiries, Gillenwater simply referred the dispatcher to the press release offering no details. Other first responders were more in the loop. One company official told a Raleigh County 911 dispatcher that at least twenty-eight miners were missing. At 4:44 P.M., records show, Massey officials asked Raleigh County's emergency coordinator for a helicopter to evacuate three injured miners. Finally, at 8:10 P.M., Massey announced just how serious the event was: seven dead miners and nineteen unaccounted for. The staging area for rescuers and families was at Whitesville, a scruffy, ancient mining town a few miles up State Route 3.

Governor Joe Manchin—a tall, stately looking Democrat who took the seat of West Virginia patriarch Robert Byrd in the U.S. Senate after he died in 2010—was in Florida on vacation to recoup from a tough session of the state legislature. When Manchin got a call from his communications director saying, "We may have a problem," he said he'd get back as soon as possible. Manchin was no stranger to mine disasters, having gone through several as governor and losing an uncle in the 1968 mine explosion at Farmington that killed seventy-eight miners and involved the largest loss of life in a coal mine disaster since the early 1900s.

In the small town of Beaver, not far from Beckley and about forty minutes away, Terry Ellison, a middle-aged blond woman who runs a home-based business transcribing medical records, heard about the blast at 5:30 P.M. "A friend called me on my cell phone," she said. "He had a police band radio scanner and had picked up the information." Her younger brother, Steve Harrah, forty, had been a miner for Massey for ten years, although Ellison wasn't sure where he worked. Massey moved miners around, and she said, "He didn't talk to me much about it, because he knew I didn't like him working in the mines."

Harrah had spent Easter with his wife's family and his six-year-old son. Ellison said, "He wanted a nice Easter meal because we lost our parents the previous year and it reminded him of ones we used to get. Then they ended up playing basketball until nine o'clock that night."

She scrambled to find a ride to the mine and didn't arrive until 7:30 P.M. "By then, there were a thousand people there. No one had called us. The company didn't call us. We all heard by word of mouth." She was herded into a training center at

Upper Big Branch's main entrance. "All I knew was that he was supposed to have gotten out of the mine at three P.M."

Word came fairly quickly, but the way Massey announced it seemed heartless. According to Patty Quarles, Massey officials reassured families that their loved ones were safe. "One official told me that Gary Wayne was safe," she says. Then they announced, "'If I call out your name, go over to Whitesville Fire Department and identify the body.' That's how cold it was."

Steve Harrah, Terry Ellison's brother, was one of seven miners found dead on a mantrip that was 1,500 feet from the mine mouth—the first of the dead to be identified. Two other miners in the group survived. The blast had gone off four miles from Harrah, but despite the distance, it was enough to kill him and six others. Ellison says her brother was not dismembered by the powerful blast but suffocated when it sucked the oxygen out of the shaft through which he was leaving. Ellison was told that family members would have to identify the body. If that weren't a blow enough, they were then sent on a wild-goose chase. "We went to the fire department, but he wasn't there. Then we went to the elementary school, but he wasn't there. Finally a police officer told us they had already taken him to the medical examiner's office in Charleston."

Another family that didn't have to wait long to know the fate of their loved one were relatives of Benny R. Willingham of Corinne, West Virginia. Sixty-one-year-old Willingham had been a Vietnam War veteran of the air force and had been a miner for more than thirty years, including seventeen years with Massey. A church deacon, Willingham enjoyed playing with his grandchildren and lifting weights. He had intended to retire five weeks after he went to work at UBB on April 5 and had reserved a room on a cruise ship with his wife, Edith Mae, to explore Caribbean islands that May. He was on the mantrip with the group that Timothy Blake had tried to rescue. Reached at her home in the tiny town of Corinne eighteen months later, Edith Mae Willingham was too upset to talk about her husband's death.

For other families, however, the ordeal of not knowing was just beginning. In Tommy Davis's home village of Dawes, tucked beside the West Virginia Turnpike that climbs high above what had been a classic coal-company town, his eighteen-year-old son, Jeff, was picking up a little brother at a school bus stop. He went back home and fell asleep until the phone rang at 5 P.M. and his mother told him that there had been a mine explosion.

The family drove over the mountain to UBB, a trip that takes about thirty or forty-five minutes. They arrived about 7 P.M. Massey put them in a building and had drinks and food for them. "We stayed up all night and at seven A.M. Tuesday, Governor Manchin brought the whole family into a room and informed us of the deaths." Manchin described telling the Davis family about Timmy, Cory, and Josh as the "most excruciating moment." He had known the extended family personally. Tommy had just one question, Manchin recalled. "'Were they all together?' I said, yes, they were."

Two days after the blast, fourteen bodies had been recovered and identified, but eleven others were found burned and mutilated beyond recognition. Another four miners were unaccounted for. Hopes rose that they might have made it to one of two airtight underground rescue chambers stocked with enough air, water, and food to last four days. Survival in those harrowing conditions was possible. In 2006, a blast at the Sago Mine, owned by International Coal Group near Buckhannon, West Virginia, killed twelve miners—but one miner, Randal McCloy Jr., was found alive after being trapped for more than forty hours despite high carbon monoxide levels.

UBB rescuers reached one of two chambers and found it empty. Unsafe levels of methane gas and carbon monoxide prevented them from going farther. On the surface, bulldozers began ripping out an access road on a steep mountain so three shafts could be drilled one thousand feet down to insert nitrogen and release enough toxic gas to let the rescue teams proceed. Families kept vigil at a Baptist church in Whitesville and at a training building at the UBB mine. State troopers in forest green uniforms helped with car rides, beverages, and food. Jim and Larry Chapman, whose brother Kenneth was one of the victims and was unaccounted for for days, say they weren't sure what to think. "The firm didn't find him but they wouldn't tell us. I knew he was dead," Jim said. His brother Larry added, "It looked like the top of that mountain was going to blow." Still, families kept vigil. One retired miner, attached to an oxygen tank because he suffered from pneumoconiosis, or black lung disease, from breathing coal dust, sat patiently in the front seat of a car waiting for word of his grandson trapped below. He was praying.

Although it wasn't apparent at the time, the blast would spell the end of the tumultuous eighteen-year reign of Don Blankenship as the head of Massey Energy. Blankenship, Gillenwater, corporate counsel Shane Harvey, and other Massey executives huddled to prepare for the inevitable barrage of bad press, shareholder lawsuits, legal challenges by dead miners' families, and massive reviews by state and federal mine-safety regulators as well as a criminal probe by the U.S. Justice Department's office in Charleston.

Under Blankenship, Massey had mastered a tactic of rebuffing almost every challenge. If MSHA issued a violation or closed a mine, Massey's lawyers were quick to sue or file regulatory appeals to blunt as many moves against the company as possible. If the Environmental Protection Agency went after one of the firm's mountaintop-removal surface mines, lawyers stood at the ready with lawsuits. The tactic was popular with political conservatives who especially admired the way Blankenship, a major donor to the Republican Party, pushed back against what they saw as a dangerous, oversized government with unneeded, profit-draining regulations, as well as against destructive, wrongheaded environmentalists.

During the five days when the fate of the missing four miners was unclear, Blankenship could be seen walking around the Upper Big Branch buildings. He issued public statements to the media and to families but kept to himself. He later told a U.S. Senate panel on May 20, 2010, that he "looked into the eyes" of family members of dead miners and said, "I don't want to do that again." He also said, "Let me state for the record: Massey does not place profits over safety. We never have, and never will. Period." Robert Byrd, who was a member of the panel at the time, responded by saying, "It is clear—I mean, clear as a noonday sun in a cloudless sky—a clear record of blatant disregard for the welfare and safety of Massey miners. Shame!"

As the chaos immediately after the blast settled, Massey officials plotted a defensive strategy that would keep investigators at bay by limiting access to employees and records. A number of miners scheduled for interviews didn't show up at first. Typically, manuscripts of their testimony would be made public, but they were kept out of public view by the U.S. Attorney's Office that was conducting a criminal probe.

Blankenship balked at appearing at investigations, although he did make it to the Senate panel on Capitol Hill where Robert Byrd berated him. In public statements, he tried to turn the tide against MSHA, claiming that the agency was embroiled in an "MSHA-gate" to cover up its mistakes and sloppiness, including its insistence on altering Massey's ventilation plans at Upper Big Branch and helping set up the mine for disaster.

There were bigger issues for Blankenship, however. For years, his tough-guy CEO stance put him at odds with public opinion and the investment community, which was used to smooth, well-managed chief executives who were sleek enough to blunt criticism, even when their firms did very bad things. Yet Massey's board of directors continued to back him even if it quietly asked him from time to time to tone down his in-your-face hillbilly posture.

As soon as the situation started to settle at UBB, questions were raised about whether this was the last straw for Blankenship. Plans were drawn up to stash enough cash to cover the UBB disaster. The company put the cost at $129 million. Blankenship was quizzed by industry analysts at conference calls if the number might be larger, perhaps $150 million to $200 million. They had good reason to do so. In 2008, Massey agreed to pay $2.5 million in a criminal fine and $1.7 million in civil penalties for a fire at its Aracoma Coal subsidiary that killed two miners in 2006. The firm had also paid $20 million, then the largest fine ever, for pollution issues related to a surface mine in Kentucky in 1999, when an abandoned mine shaft under a pond holding mine waste split open and drained the waste into the watershed.

Immediately after the explosion, Massey's stock tanked

from the mid forty-dollars-a-share level to the mid to low thirty-dollars-a-share range, setting it up for a takeover. Its reserves of metallurgical coal as well as some of the top-rated steam coal for electric utilities made it a prize to covet. Also, the year before, Massey had announced the $960 million purchase of Cumberland Resources, also of Abingdon, a purchase that at the time doubled Massey's sales of metallurgical coal. Alpha Natural Resources, which is based in Bristol, Virginia, and had ample reserves in Central Appalachia plus easy-to-mine, low-sulfur coal in the Powder River Basin of Wyoming, had already made a run at it. As stock prices fluctuated, the coal industry was in the midst of a round of restructuring, and Massey was in the crosshairs of competing firms. And as expected, Alpha Natural Resources was preparing another run at Massey. Other interested parties included ArcelorMittal, a big Luxembourg-based steelmaker that got its start in India, as well as U.S. coal giant Peabody.

Blankenship was dead set against a takeover since it would cost him his job. The stakes were high. He had to manage the aftermath of the worst coal mining disaster in forty years in the United States and convince his directors to hang tough against a takeover.

Back at UBB, several heartbreaking days passed. Four miners remained unaccounted for. Families of miners whose bodies had been recovered and identified were busy with funerals.

Finally, at 12:30 A.M. on Saturday, April 10, Manchin brought the media, which by now included representatives of outlets from around the world, to the Marsh Fork Elementary School and announced: "We did not receive the miracle we prayed for."

The announcement was especially bitter because a federal law passed after the 2006 disaster at the Sago Mine killed twelve miners required faster response times to mining accidents. In that incident, the media prematurely released news that the twelve men had been found safe and then had to correct themselves. At UBB, MSHA started a massive probe, but it was weeks before the toxic gases had cleared enough for them to have complete access to the mine. Towns such as Whitesville returned to their sleepy selves as a handful of families talked to their lawyers about suing Massey. Eventually, twenty-nine wrongful death lawsuits would be filed by families of the victims, along with a host of shareholders' lawsuits from institutional investors such as the Macomb County Employees Retirement System.

Massey officials got to work figuring payments to the families. For a time, Massey offered each family $3 million, but the offer expired when it was taken over by Alpha Natural Resources. To its credit, the company kept paying the salaries of surviving miners even if they didn't report to work. That was the case for Tommy Davis, who stayed out of work for more than ten months but kept getting a paycheck from Massey. His house and yard are made over as a memorial to his lost relatives. A Ford F-150 pickup truck in the driveway has a big decal on its rear cab window commemorating his son, Cory. The front door of his home has the following inscription:

> *In loving memory*
> *My hero*
> *My best friend*
> *My son*

Inside the living room there is a stack of taupe-colored Massey Energy work shirts. The family keeps them on hand to fly from their flagpole in the front yard, just under the U.S. flag, as a tribute to Cory. Tommy says he faults Massey for not telling the truth faster just after the blast, but "what's done is done." Eventually, he decided to go back to work at another Massey deep mine, but first he was going on a road trip to Daytona Beach, Florida, on his Harley-Davidson: "I need to face down my fears and deal with my demons. Get my mind right."

2

RENEGADE CEO

In person, Donald Blankenship can be surprisingly under-whelming. When I first met him at the Richmond, Virginia, headquarters of Massey Energy in the fall of 2002, he fixed his brown eyes on me as I asked my questions and made my pitch to visit one of his coal mines with a photographer. His eyes were steady and unblinking, almost expressionless. Deep, fleshy jowls and his square face gave him the appearance of an oversized basset hound. He answered my questions in a few, carefully chosen words, as if to reveal only enough to keep our meeting moving along. There was no hype, no anima-tion, and no self-aggrandizement in which some corporate chief executives are prone to indulge. While it was far from the most revealing interview I'd ever had, I got my wish, and a few weeks later, a photographer and I were a little over five miles inside a mountain near the small town of Red Ash in Southwest Virginia.

Blankenship's forgettable presence belies the internal power that had made him one of the most remarkable, if not vilified, chief executives of any major American corporation in recent memory. The scrolls of dead miners who worked for him and the massive ecological devastation Massey has caused Central Appalachia's ancient mountains seem in marked contrast to Blankenship's outwardly serene manner. Yet Blankenship, who declined to be interviewed for this book, is known as an outrageous micromanager who can pitch a fit if a coat is hung up the wrong way in a closet. He was a stubborn taskmaster who wouldn't budge from production goals and demanded complete obedience as he sought to attain them.

Blankenship's character is marked by bluntness and loyalty. Never afraid to mince the words that he speaks sparingly, Blankenship is as consistent as he is contrarian. One may hate what he has to say and bristle at the boorish way he may say it, but he doesn't lie, at least not by his own standards. He may go way over the top by refusing to admit to the most obvious safety problems, and his mania for cost cutting may risk lives, but he's consistent. Friends and enemies alike admire his uncanny knack for numbers and for using accounting to understand the larger business picture. And he takes fierce pride in a small tract of rugged mountains on the Kentucky–West Virginia border where he was born, raised, and still maintains at least three expensive homes on either side of the Tug Fork in an area that is one of the poorest and least healthy in the country.

In coal circles, Blankenship was seen as an annoying and frightening anachronism, a lightning rod that drew unwanted attention and criticism when coal operators were having trouble enough promoting a filthy and deadly, if not dying, industry. His cult-of-personality style of management was also unique in a field where many coal companies are run tightly by professional investor relations specialists, protective human resources bureaucrats, or private equity suits. According to Joshua Frankel, manager of pricing and structuring at Duke Energy Corporation in Cincinnati, Massey Energy is the last of the old-style coal operators. "Massey was that old-style model of 'balls of steel' guys. The old school was a seat-of-your-pants—you counted your fingers after you shook hands with them. Now, it's a brave new world. You just don't run down and get permits anymore. The attitude is we'll sell coal on a global market wherever we get the most amount of money. The coal industry is going through a transition, and you need scale and capital."

Bruce Stanley, a lawyer with the Pittsburgh office of Reed Smith who also grew up in Blankenship's home of Mingo County, West Virginia, and has fought him in court for years, also describes him as the last of a breed. "He's robber baron, more comfortable in a boardroom in Pittsburgh in the 1890s. But Mingo County is still like 1890. He's left a lot of scars." Stanley adds: "He's clearly a resolute kind of guy in what he is trying to do. He has a divide-and-conquer mentality and has turned people in a community against each other. His legacy is a Mingo County with its coal ripped out, its mountains destroyed, and the people without jobs. He's walked away with a quarter of a billion."

Blankenship's story is one of rags to riches. He was an illegitimate child, the product of an affair his mother had with a man while her husband was in military service in Korea. His mother, Nancy, had three other children—Anthony, ten; Beulah, eight; and George, five—when Don was born in a small building that is now the site of a gasoline station in Stopover, Kentucky, in March 1950. Stopover is little more than a crossroads with the small Church of God with a sign STOP INSIDE FOR A FAITH LIFT and downhill from a small cemetery with a sign stating D. H. BLANKENSHIP and the year 1897.

Blankenship's ancestors are believed to have come from Germany and England sometime in the 1700s, and Nancy was a descendant of the McCoy family famous for its late-nineteenth-century feud with the Hatfields that has long been celebrated in history books, movies, and cartoons.

Nancy and her husband, Anthony, later separated, and Don and his brother George ended up in Virginia for about a year before she opened a store in Delorme, also known as Edgarton, just across the narrow Tug Fork from Kentucky on the West Virginia line. She took money from her divorce settlement for the store. Conveniently located just in West Virginia, which allowed beer sales, it drew thirsty customers

from dry Kentucky along with the Mountain State. Delorme was also on the main line of the Norfolk and Western Railway, a major coal hauler that served many mines in the region. In the 1950s, the N&W's crack Powhatan Arrow passenger train with its streamlined steam locomotive and classy maroon, yellow, and black cars would roar past.

Nancy labored at her store seven days a week, and her children practically lived there. For dinner, she'd make a pot of beans or potatoes on a stove as she tended to her customers. Don found himself following her example. She was quiet but savvy and was very blunt and quick to let her opinions be known. She was a conservative Republican who backed Barry Goldwater in 1964 and despised President Lyndon B. Johnson's Great Society programs to help the poor with what she considered unfair and unhealthy welfare-state policies. Don's brother Anthony told the *Charleston* (West Virginia) *Daily Mail* in July 13th 2005 article that she would point out drunks and deadbeats in the tiny world of Delorme as examples of what not to be. On payday, miners congregating in Delorme hit the beer and often got into fights, some with knives. That drew the scorn of Nancy, who nonetheless made a living keeping them supplied with brew.

Constant work at the cash register honed Blankenship's natural gift for math. In the words of Morgan Massey, the former coal company president who hired him in 1982, "Don can tell you what this quarter's numbers are. He can also tell you what next quarter's numbers will be," he adds, noting that Blankenship could somehow factor unknowns into his estimates, including accidents and shutdowns that couldn't be foreseen. For high school, Blankenship traveled a few miles up windy State Route 49 to Matewan, a small town but bigger than Delorme. He graduated second in his class. While he wasn't known for playing any sports in particular, he did pick up an interest in collecting baseball cards along the way and is said to have an extensive collection—his fantastic talent for numbers extending into his hobbies. In 1968, he enrolled at Marshall University in Huntington, West Virginia, to study accounting. Short of funds, he undertook a heavy class load to graduate in three years, taking time off between his second and third year to work a union job as a Pittston coal miner, where he handled explosives, ran a shearing machine, and installed belts. After graduation, he found a job in Chattanooga as an entry-level accountant, where he spotted an advertisement for Keebler, the cracker and cookie baking company. Keebler took him to Macon, Georgia; Chicago; and Denver, where he married and had two children, John and Jennifer. He was divorced in 1992. Following that, he worked in accounting for Flowers, another bakery, in Thomasville, Georgia.

By that time, he was on Massey's radar screen. Massey, then known as A. T. Massey, operated a number of mines in Blankenship's home area, stretching from Williamson, West Virginia, a major coal supply town and rail yard, to the Matewan area. It had scouts about sniffing for talent and offered

THUNDER ON THE MOUNTAIN

Blankenship a job in 1977. He turned the company down. After he quit his second job, a second offer from Massey came, and he took it, landing him at Rawl Sales & Processing Company, just a little west of Matewan and his circle of friends. The area is so precious to Blankenship that years later he would build one of his mansions there—a rambling, Victorian-style house painted turquoise and white, set back on some rolling hills and shaded by magnolia trees. It is surrounded by a black metal fence with an electronic driveway gate.

At the Rawl job, Blankenship got to work crunching numbers, and it was there that he had an experience that would affect him profoundly for the rest of his career. By 1984, he had risen to running the Rawl mining operation and got caught in a strike precipitated by E. Morgan Massey, who had never loved unions and saw an opening to break the United Mine Workers of America. Its president, Richard Trumka, now head of the AFL–CIO, allowed some UMWA units to negotiate separately with coal operators rather than go through the umbrella management group called the Bituminous Coal Operators Association. Massey moved in quickly to try to get the UMWA to negotiate as many as fourteen different contracts with the union rather than one omnibus agreement, touching off a strike. Massey management had posted private security guards throughout its operations, including at Rawl coal. Accusations of provocations and violence erupted from both sides. To keep Rawl running, Blankenship brought in

38

nonunion miners from Kentucky to crash picket lines. Union sympathizers retaliated, and Blankenship said that three of his armor-plated cars were shot up during the fifteen-month-long strike. At one point, drive-by shooters sprayed his office with bullets. Blankenship has said that eleven rounds penetrated the building and smashed the screen of an old Zenith television set that he still keeps as a souvenir of the violence.

The incident radicalized him. He had never had any sympathy for the UMWA, but his attitude turned from disdain to hatred. It was to color his attitudes as the union repeatedly tried to organize Massey mines, which to this day remain the least unionized in the United States. Blankenship's hardnosed conservatism, which he learned at his mother's skirts next to a convenience store cash register back in Delorme, went to a whole new level of defiance. His in-your-face persona was born.

When Blankenship was hired at A. T. Massey Energy, the firm was owned by Fluor Corporation, the global engineering giant that was based in Los Angeles and specialized in petroleum and petrochemical production, besides being a major defense contractor. In 1981, it bought St. Joe Minerals, which had bought A. T. Massey in the 1970s along with lead and zinc properties. Throughout the 1980s, E. Morgan Massey served on the Fluor board. But Massey found the coast-to-coast trips tiresome and felt he never got the attention at board meetings that his company was due, even though it provided Fluor with more than $1 billion, or roughly one-third, of its revenues.

Massey noticed Blankenship's accounting talents and his hard-nosed attitude fit the bootstrap history of Massey's ancestors, who arrived in West Virginia from Wales in 1876 and built a coal empire. Blankenship's bluntness also was a plus. "Don always tells you the truth, whether you want to hear it or not." Blankenship also shared Massey's vision of beefing up the company's stocks of high-quality, low-sulfur coal that could go either way in the market. It fetched the highest prices from utilities for a fuel that met tougher air-pollution standards. It also had properties that were ideal for making coke for smelting steel. Although the U.S. steel industry was in serious decline because its ossified managements were too cheap to upgrade equipment, steel production was booming in places such as South Korea, Japan, Brazil, and Germany, and would soon be in China as well.

They were all hungry markets waiting to expand. In fact, it was the quest for more and better coal reserves, especially met coal, that eventually got Blankenship started on his political career. By 1991, E. Morgan Massey had all but retired from A. T. Massey, leaving Blankenship as its CEO and giving him more influence with Fluor. Two years later, a small coal operator named Hugh M. Caperton spotted a beat-up mine in Buchanan County in Southwest Virginia not far from his West Virginia home. The mine had high-quality metallurgical reserves, so Caperton bought it.

Moving quickly, he fired the contract miners who had operated the facility, replacing them with 150 more professional and unionized workers. In less than a year, the mine had quadrupled its output to 1 million tons annually. That caught the interest of Blankenship, who operated Massey mines in the same area. He made an offer to buy the mine. Caperton refused, but rather than abide by the unwritten "live and let live" code of coal operators, Blankenship threatened him. "He said, 'Don't take me to court. We spend a million dollars a month on lawyers, and we'll tie you up for years.'"

When Caperton held firm and still refused to sell his coal mine, Blankenship swooped in and bought up land around the Caperton property, making it nearly impossible for him to move coal by truck or rail. Blankenship then lured away Caperton's primary customer. The skirmishing went on for four years before Caperton finally caved in and agreed to sell. But the battle was just beginning. According to the *American Bar Association Journal,* on the very day the sale was due to close, Massey walked away from the agreement, forcing Caperton's firm into bankruptcy.

Fighting back, Caperton sued Massey for fraud and interfering with a contract, winning a $50 million jury judgment. Blankenship appealed, and in 2004 he made his leap into politics. In West Virginia, judges are elected, not appointed, including those on the state supreme court of appeals. Blankenship realized that clout on the court was important because of his Caperton appeal and various other legal issues. So he pumped $3 million into the campaign of a little-known Charleston lawyer named Brent D. Benjamin for a twelve-year term as a state supreme court justice. In August 2004, Blankenship formed a so-called section 527 fund-raising organization, titled And for the Sake of the Kids. In reality, the 527 fund, named for an Internal Revenue Service code, was a ruse that not only supported Benjamin but also tarnished his opponent, Democratic incumbent Warren McGraw, who had generated negative publicity for helping release a child molester from prison. According to the *ABA Journal,* Blankenship also thought that McGraw's rulings were "antibusiness."

Blankenship contributed $2.4 million for the fund's war chest, which totaled $3.6 million. Among other things, it was used to spread billboards around the state asking, WHO IS BRENT BENJAMIN? Blankenship thought up the billboard content on his own and joked he might be in the wrong business. The campaign worked: Benjamin won with 53 percent of the vote in November 2004. Three years later, Massey's appeal of the Caperton decision was granted by a 3–2 state supreme court vote, with Benjamin voting with the majority. According to the *ABA*, "Benjamin never acknowledged Caperton's disqualification motions. Caperton's lawyers never got to argue them orally or received an explanation for the decision."

In June 2009, in a stunning slap, the U.S. Supreme Court

declared that Benjamin had denied due process by refusing to recuse himself in the Massey case. A new suit has been filed in Virginia. If anything, the case shows just how contentious and tireless Blankenship could be when he chose to deploy his engulf-and-devour tactics. When he was in business, he wanted to make it crystal clear that he would spend any amount and take as many years as he felt necessary to achieve his goals.

By now, Blankenship was a major voice in West Virginia politics, making some wonder if he might run for office. He was everywhere, suing Governor Joe Manchin for violating his First Amendment rights by threatening to have the state scrutinize Massey's mines more closely because of Blankenship's political activities. Blankenship sued the *Charleston Gazette* for its aggressive investigative reporting of him and hit the rubber chicken circuit, decrying "greeniacs" and "excessive" federal regulations. Blankenship dropped his suit against Manchin after Manchin said that he regretted some of his remarks. The newspaper suit was dismissed. Blankenship donated thousands of dollars to help the state Republican Party build a new headquarters. According to former Matewan mayor Johnny Fullen, "I was in their office one day and he got a call from the German government. One day Karl Rove called. Another time, he had just gotten back from a meeting with Henry Kissinger." In Richmond, however, the

Massey board was getting uneasy about just how much publicity its chief executive was kicking up, but any measures they took to cool Don down didn't seem to get anywhere. If anything, Blankenship was growing more outrageous by the month.

In the summer of 2006, it was later revealed, Blankenship and a female friend traveled to the Mediterranean coasts of Italy, Monaco, and France to take in the sights and the cuisine. Accompanying him was his friend, Elliot "Spike" Maynard, who also brought along a female companion. Maynard was an old friend of Blankenship's who had grown up not far from Matewan. But it was Maynard's job that made the trip so controversial. Maynard was chief justice of the West Virginia Supreme Court of Appeals and was in a very strong position to influence the cases that Massey brought to court.

When pictures of the two men were published in state newspapers in early 2008 showing them smiling in leisure shirts on a seaside balcony on the French Riviera, many were outraged. Maynard claimed he was just a friend of Blankenship's who happened to meet him overseas. Sure, they had meals together, but they'd alternate who paid for dinner. "We have dinner in West Virginia, too," Maynard explained. Maynard, however, did withdraw from a case involving Massey after the trip was exposed and was later beaten at the polls in a primary. The association also put a permanent cloud over Blankenship as one who happily wielded influence without giving a hoot for how badly it tarnished his public image.

Previous to that, Blankenship's imperious behavior landed him in a lawsuit with his maid, Deborah May. In a case that continues to generate publicity, May sued Blankenship, claiming that he forced her to quit her job in November 2005. Among her alleged faults: mixing up a takeout order from McDonald's, misplacing ice cream in a freezer, and improperly hanging a coat in a closet. May also claimed that Blankenship kept on adding to her duties, such as requiring her to clean up the interior of a recreational vehicle his son used when he raced late-model cars on the dirt-track circuit. May sought unemployment compensation and won her case with the state supreme court in 2008. This time, the court came down on Blankenship, saying that he "had physically grabbed her" and that his "shocking conduct" was "reminiscent of slavery and is an affront to common decency." In court documents, Blankenship claimed that May had filed the suit to get a higher salary.

Blankenship's personality also had a big impact on his coal-firm employees. When freshly hired middle managers took their desks, they'd usually find a can of Dad's root beer, a regional brand, on their desks. Confused, they would ask about it and be told that it stands for "Do as Don Says," explains lawyer Bruce Stanley. By about 2005, Blankenship was reaching the apogee of his erratic, power-driven behavior. His early successes with accounting, coupled with a stubbornness he acquired during the union wars of the 1980s, had transformed him into an egocentric bully who flourished by playing politics and abusing his employees. He may not have realized it, but he was setting himself up for a big fall.

Travel to Matewan, however, and the impression of Blankenship turns 180 degrees. The small town of about five hundred with its redbrick buildings is an historic landmark. A clapped-out pickup truck sports a bumper decal stating: "Some Preachers Lie. God Knows Who." The community was where union activists, aided by town police chief Sid Hatfield, fended off early-day security thugs from a notorious firm called the Baldwin-Felts Detective Agency that operated out of Bluefield, West Virginia, and Roanoke, Virginia. As portrayed in the 1987 John Sayles movie *Matewan,* Hatfield, played by actor David Strathairn, backs his people when they are threatened in 1920 with firing by the Stone Mountain Coal Company if they unionized. They would be replaced by scab laborers such as African Americans from the Deep South and newly arrived Italian immigrants. The movie's climax is a bloody shoot-out on Matewan's Main Street between Hatfield, his deputies, and the hired gunslingers working for the coal firm. Although much of the movie was actually filmed in Thurmond, another historic West Virginia coal town, Sayles's movie transformed Matewan into a tourist destination, and it did so with the help of Don Blankenship.

At the exact spot of the Hatfield shootings is a gray state

historic marker in the parking lot of a post office. Abutting the lot is the Historic Matewan House Bed and Breakfast, a restored building with heart-of-pine floors and an impressive stone fireplace. It is operated by Pat Garland and his wife. Garland, who is a retired Southern Baptist missionary originally from Ohio and died on October 28, 2011, after a long illness, was a friend of Blankenship, who occasionally drives up from one of his houses in the area to visit. Blankenship is said to own an impressive collection of cars, including a Lamborghini, a Porsche, a vintage BMW, and a 1957 Chevrolet. One day, Blankenship drove up to the bed-and-breakfast in a new Bentley and said, according to Garland, "Come on in and take a ride." Garland said he replied, "I had never seen a car like that and might slit open a leather seat with a screwdriver in my pocket."

Garland, who is not shy about asking his hotel guests if they have accepted Jesus Christ as their personal savior, frequently quotes the Bible when describing how the world is misjudging Blankenship. Recalling the story of David and Goliath from the Old Testament, Garland envisions Blankenship as a misunderstood underdog facing down evil and danger. "Don saw great needs. He's a great seer. A bunch of people don't see that he is a man with a big heart." Blankenship did hand out frozen turkeys and bags of potatoes to the poor at Thanksgiving and Christmas. "His name and influence will be here for a long time," Garland says.

One humid spring evening, Garland is flipping burgers

down the street from his bed-and-breakfast for the all-class reunion of Matewan High School. Old and young couples mingle next to a huge concrete flood wall that shields the town when the otherwise tame Tug Fork roars to life with occasional floods caused, in part, by the mountaintop-removal practices of firms such as Massey Energy. A local band pounds out country and rock oldies. Garland expected Blankenship, class of 1968, to show up for the reunion, but he did not.

Next to the celebration is a monument to Blankenship, built partly with his help. The monument is the Matewan Depot Replica and Museum, an attractive redbrick building remade to look like an old Norfolk and Western Railway station. The tidy museum has plenty of exhibits. There are old newspaper advertisements of showings at the local movie theater in January 1950, including John Garfield in *Force of Evil* and Gary Cooper in *Pride of the Yankees*. An HO-gauge model train set shows the famed N&W Powhatan Arrow passenger train whipping around little mountains. There are plenty of period photos depicting different types of coal mining. One picture is an artist's rendition of how a mountaintop-removal surface mine could be remade into a car racetrack with bleachers and concession stands. In fact, a new regional high school a few miles away is sited atop a converted strip mine. And there's a large, formal photographic portrait of Don Blankenship staring at the viewer without a smile, along with information stating all the good things that he and

Massey Energy have done for the community. According to local accounts, Blankenship sank about $750,000 of his own money into the Matewan museum and other community improvements.

Blankenship's closest local collaborator lives almost in the middle of Matewan, in a modest white house between Garland's hotel and the museum: Johnny Fullen, an African American man who was town mayor from 1984 to 1998 and works in the coal business. Fullen invites guests into his living room, where a thick book about the early days of the U.S. Constitution by Pauline Maier, a history professor at the Massachusetts Institute of Technology, along with a number of magazines, lies on a coffee table. Fullen says he gets the Sunday *New York Times* but is annoyed with the newspaper because he gave a *Times* reporter a two-hour interview about Blankenship and "nothing in the article was right."

Blankenship is regarded by some as a kind and thoughtful man who has limitless goodwill for the people of the area. "He calls me John," says Fullen. "He says he thinks that people don't think he wants them to have a two-car garage and a house, but that's what I want for them, Don would say." Indeed, he adds, Blankenship has been consistently the go-to guy for help when something goes wrong in the community. Fire, flood, or illness can all bring Blankenship to stroke a check and put things right for his hometown people. If a ball field was ravaged by floods or the hungry needed Thanksgiving turkeys, Blankenship came through.

Fullen has known Blankenship and his family for years, but what got Fullen intimately involved with Blankenship dates back to the days when movie director John Sayles was negotiating with the town of Matewan for access for his film. Sayles had sent a representative to the town residents, who were worried that the movie would portray them as dumb hillbillies. Sayles wanted to reassure them that would not be the case, Fullen says, but he ended up doing most of his shooting in Thurmond and other areas for other reasons. Sayles's pitch, however, gave the town's leadership an idea: If Matewan is so historic that a major director is willing to name a movie after it, maybe they could find lucrative opportunities exploiting that history themselves.

So they formed the Matewan Development Center with help from the Matewan National Bank, Massey Energy, and other coal operators. The idea was to raise money to renovate the town's several historic buildings, some with notable turn-of-the-century architectural flairs. "Our first goal was $150,000," says Fullen. The money would be used to help get the town onto the National Register of Historic Places, which would open the door to tax credits and more renovation money. The state came up with $500,000 to help restore the R. W. Buskirk Building downtown and push ahead with a makeover for other buildings. But the plan took a big hit when the Buskirk building burned down and planners found themselves fifty thousand dollars short of a financial goal. Fullen says it was a low moment as he stood on Matewan's Main Street and pondered the problems. Then he noticed Blankenship. "Don was walking down the street and said, 'What can I do to help?' I told him we would need more money and he just said, 'What do you need?'" Blankenship told Fullen to come to his office, and when he did, he was given a check for fifty thousand.

Blankenship became Matewan's municipal Santa Claus. Whenever the town needed something, he was there, Fullen says. If they needed a new school, Blankenship kicked in ten thousand dollars. If they wanted a new baseball field for the Little League that cost ninety thousand dollars, Blankenship would pay fifteen thousand of his own money and send over some Massey construction teams to build it. When the Tug Fork churned up with flooding waters and ruined the high school's football field, Blankenship quietly sent over a check, adding money for flood-wall improvements. His largesse went beyond town limits to payments for medical care for the needy at a hospital and clinics in Williamson, West Virginia, near Matewan. Massey also underwrote scholarship funds and helped with capital investment at schools ranging from the Williamson Middle School to the University of Kentucky. Its Doctors for Our Communities program with Blankenship's alma mater of Marshall University provides tuition loans to incoming medical students and then forgives those loans if the newly minted doctors stay for seven years in the medically underserved area where Massey operates. West Virginia University and West Virginia Tech got $500,000 for their mine engineering programs.

Nowhere, however, is Blankenship's generosity more evident than in his hometown of Matewan, which has succeeded in turning itself from a forgotten railroad stopover into a tourism destination. Several hotels open during warm weather months to accommodate riders who roar about the woods in all-terrain vehicles to follow the Hatfield-McCoy Trails, a system of marked trails running through several hundred miles of Mingo and surrounding counties, sometimes on reclaimed land from surface coal mines. Matewan jealously keeps and markets its down-home atmosphere. A downtown restaurant, the Matewan Depot, thumbs its nose at cholesterol and plays hillbilly by offering a menu with deep-fried Oreo cookies and monkey bread. A nearby tavern is outfitted in rough and rustic wood panels.

Just after the explosion at Upper Big Branch, Blankenship kept to himself as he appeared at disaster centers near Whitesville, where families of the dead and missing had gathered. Other, more people-savvy individuals, such as then governor Manchin, hugged and consoled coal families he had known for years. At times, Blankenship's tactlessness seemed hurtful. Miner Gary Quarles and his wife, who lost their son at Upper

Big Branch, were stunned when Blankenship gave one of their grandchildren, the child of their dead son, his business card at a crowded memorial service attended by President Barack Obama. Blankenship told the boy, "Hang on to that. Not many people get one of those." His statement showed such over-the-top self-absorption that the Quarleses keep a photograph of Blankenship handing over the card to their grandchild on the screen saver of their home computer. "We don't want to forget this," Quarles says.

Blankenship stubbornly tried to hold on to his job and keep Massey independent. For some months, he continued his bluster. Within two weeks of the mine blast, Massey Energy had hired Public Strategies, an Austin, Texas, public relations firm, to handle crisis management. The firm was a natural. Aides to President George W. Bush, Mark McKinnon and Dan Bartlett, as well as Jeff Eller, a onetime aide to President Bill Clinton, staff it. Conveniently, Public Strategies is headquartered in Austin, a few blocks from the office of Bobby Ray Inman, then lead independent director of Massey. Public Strategies' ploy was tailor-made for Blankenship: Admit nothing, insist coal dust and faulty equipment had nothing to do with the blast, and try to make federal regulators, especially the Mine Safety and Health Administration, the bad guys. At the Bluefield Coal Symposium in September 2010, for instance, Blankenship was comparing MSHA's probe of Upper Big Branch to Watergate, claiming the federal government was lying and covering up its errors. He also managed a pitch for his old judicial friend from the French Riviera trip, Elliot "Spike" Maynard, who was running unsuccessfully for Congress as a Republican against longtime incumbent Nick Rahall.

It was clear, however, that his position was untenable in the wake of the most fatal mine explosion in forty years. People were tired of Don Blankenship, and his antics came back to haunt him. Few had any sympathy for him, outside of a handful of folks back in Matewan. Blankenship's fight to save his job and keep Massey independent was futile. Several coal firms, most notably Arch and Alpha Natural Resources, plus some Indian firms, badly wanted Massey's rich reserves of metallurgical coal. The downside was they would have to deal with Massey's lawsuits and reshape its safety program, corporate culture, and bad corporate name. Still, Massey was worth it, and Alpha finally won the fight—and on December 3, 2010, Blankenship was gone, officially "retired" at year-end. He had declined to testify before several state and federal investigators and still may not be done as far as the civil and criminal consequences of his management of Massey Energy.

He didn't have to worry about his income. He received a severance package valued at $86 million, an option to buy land at the turquoise house near Matewan, and title to a 1965 blue Chevrolet truck that he had previously transferred to the company. But most important, Blankenship will get access to company lawyers at company expense if he is sued or investigated.

After his forced retirement, Blankenship vanished. A lawyer says he has been sending legal materials to Blankenship at an apartment address in Johnson City, Tennessee, a small city a couple of hours away. In his new location, he is close to his son John, who races in the Lucas Oil Late Model Dirt Series for automobiles. Using a car shop in Morristown, Tennessee, as his base, the younger Blankenship and his entourage, now including his father, spend their weekends competing on the dirt track circuit, heading to spots such as Rome, Georgia; Winchester, Virginia; or the Atomic Speedway in Alma, Ohio. Mullen says that Don Blankenship has invited him several times to come to the races and that his son's performance has improved with his assistance.

Back near Matewan, the turquoise house still appears well maintained, as is another turreted mansion he had built atop a mountain on the Kentucky side of the Tug Fork Valley. Local Matewan people said he would show up on occasion, but those occasions were rare.

Six months later, at Marriott's MeadowView Conference Resort and Convention Center in Kingsport, Tennessee, a few dark-suited Massey top managers milled about a conference room as the proxy votes from Alpha and Massey shareholders were counted before the companies were formally merged and the Massey name and stock ticker (MEE) officially disappeared. Some Massey executives still worried whether they'd be hired by the new firm.

While Don Blankenship was nowhere to be seen, his legacy

will last for years. Stanley says he has been deposed in a now-settled lawsuit with Massey representing seven hundred people who claim that Massey pumped from 1.1 billion to 1.4 billion gallons of toxic sludge into abandoned underground mine shafts, spoiling their drinking water. That's at least five times the amount of pollution as in the 2010 *Deepwater Horizon* explosion and oil spill in the Gulf of Mexico, notes Stanley. Alpha Natural Resources agreed to settle the case in July 2011, just a day before the trial was to begin. That signaled the new owner's strategy: Clean up as many of Blankenship's legal messes as quickly as possible and then get on with the extremely lucrative business of mining.

In the aftermath, Blankenship "has left a lot of scars," Stanley says. It could be that Blankenship wrote his best epitaph yet when he was quoted in a West Virginia Public Television biography: "I'm the one who will die with the most money."

3

UP THE HOLLOW

My introduction into the poverty of Appalachia came in late 1962, when I went to a little coal town near Clarksburg, West Virginia, to help my father distribute Sabin oral polio vaccine in sugar cubes. I was nine years old and had just moved to West Virginia from the Washington, D.C., suburbs, where Dad, a career navy doctor, had been assigned before retiring and going into private practice. The little town couldn't have been farther from affluent Bethesda, Maryland, where we had lived. Tables had been set up in a school gymnasium that was missing part of its wall. Snow was drifting onto the basketball court. None of the local people there seemed to sense anything unusual about the state of the school.

It was only one of several culture clashes I was to have at that young age. In fourth grade in Bethesda, I had been taking French. In fifth grade at Towers Elementary School in

Clarksburg, teachers told me that studying a foreign language was useless because "you're never going to use it."

I was required to take official West Virginia history in grade school. We spent a lot of time learning how to spell "rhododendron," the state flower, but huge and important parts of the Mountain State's story were left out because they were too politically potent. We weren't supposed to know about class exploitation. We never were told how outsiders cheated ignorant hill folk out of their mineral rights or about the 1921 Battle of Blair Mountain, pitting thirteen thousand striking coal miners against police and National Guardsmen armed with tanks, machine guns, and biplanes. That was the biggest civil confrontation in the country since the Civil War, but we never read about it. It wasn't until years later that I learned that famed Socialist Mary Harris Jones, aka Mother Jones, had been tried several times in Clarksburg courts for her activism on behalf of coal miners. "Medieval West Virginia" was her description of the Mountain State.

How the enormous contrasts of Central Appalachia—great riches against abject poverty—came about is rooted in brutal exploitation so intense that it is an anomaly in the history of the United States. Few other parts of the country have endured more than a century and a half of such stark contrasts. While they have moved on with new industries and new lives, in many respects West Virginia and Eastern Kentucky remain stuck in the late nineteenth century.

•   •   •

Flash forward forty-eight years. Much of the poverty and poor health and education remain. Early on a Saturday morning, photographer Scott Elmquist and I are on U.S. 119 heading from Pikeville, Kentucky, into southern West Virginia to attend a rally against mountaintop-removal surface mining. As we pass by Pike County High School on the four-lane road, I notice a sign out of the corner of my eye advertising RAM. Something clicks in my head, and I turn onto the road leading up to the plateau where the modern-looking school and a packed parking lot are situated.

RAM means Remote Area Medical Volunteer Corps—a group that travels to mostly rural and impoverished areas of the United States offering free medical and dental care. RAM was started by Stan Brock, a self-styled adventurer and bush pilot who spent years in the Amazon before returning to the United States to found the organization in 1985. Much of the year, RAM volunteers spend weekends in little coalfield towns and rural areas like Pikeville or Evansville, Tennessee. Poor folk who can't afford health insurance routinely show up in the early-morning darkness to line up for perhaps the only medical and dental checkups they'll have that year. In other words, it is as if I am back with my father in a corner of Harrison County, West Virginia, in the late fall of 1962.

More than a century of coal mining in Pike County, one

of Kentucky's biggest coal producers, apparently hasn't made much of a dent in poverty. Poverty rates for the county were 27 percent in 2009, about 10 points higher than the state average. In other words, there's not much of a trickle-down factor for ordinary Pike County residents despite the riches coal operators export from their county.

The corridors of the high school are packed with up to a thousand people, many wearing blue jeans and country-style shirts. Among them are 120 medical and dental personnel of the U.S. Public Health Service wearing navy-style combat utilities in today's digitally patterned blue and gray design and black combat boots. We ask for a public affairs officer and soon are talking with Commander Jim Mason, a trim man in his early forties, who says the Public Health Service, a federal organization tasked with helping deal with epidemics and natural disasters, is helping out RAM. It has brought in doctors, dentists, and medical assistants from as far away as Alaska and Puerto Rico to treat Eastern Kentuckians for free that weekend. "We look at this as a win-win," he says. "This is an opportunity to give back to the community and we can keep our skills sharp for a natural disaster."

If preparing for natural disasters is part of the mission, then the high school makes for an excellent place to train because it is certainly a disaster even if there isn't anything natural about it.

Classrooms are transformed into medical examining rooms. And in an eerie scene that brings back the West Virginia of my youth, the gymnasium has been transformed into a gigantic dentist's office, although in this case there's no snow wafting in from a torn-out wall. There are eighty-four dental chairs on the basketball court. Each one is filled with a person as dentists and assistants in blue surgical scrubs work at their teeth. Among the dental personnel are volunteers from the University of Louisville School of Dentistry, who also helped bring in the dental chairs. There is a line of people waiting patiently for treatment. One is Tommy Johnson, a young man from the little burg of Blackberry, Kentucky. "I've been here since one A.M.," says Johnson, who notes that his job as a security officer does not include dental insurance and he waits each year for RAM to show up to deal with his dental problems.

Despite its massive riches of coal, Central Appalachia is indeed a busy place for RAM. Travel about forty miles to another coalfield spot, Wise County, Virginia, where RAM also makes annual visits. A year or so before the Pike County visit, about three thousand people snaked along Hurricane Road at 5 A.M. on their way to the Wise County Fairgrounds. Waiting for them were eight hundred doctors, nurses, nursing assistants, dentists, and dental hygienists. Yet Wise County is a little better off than Pike County across the border. Wise County has had poverty rates of 19.2 percent, or more than twice that of the rest of Virginia. The per capita income is only fourteen thousand dollars a year. It is one of two counties of Virginia's ninety-five counties that are designated as being economically distressed by the Appalachian Regional Commission (ARC), a quasi-federal body set up after President John F. Kennedy was shocked by the poverty he found while campaigning in West Virginia in 1960. In that state, the ARC rates twelve counties as poverty-stricken, including McDowell and Mingo, which are among the largest coal producers. In Kentucky, the situation is far worse. Some 25 counties out of 125 in the state are distressed, and that includes the entire Appalachian coalfield area of Eastern Kentucky.

The data is from surveys taken from 2000 to 2004, showing that things haven't changed all that much since firebrand Kentucky lawyer Harry M. Caudill revealed to the nation just how widespread and profound Appalachian poverty was with his 1962 book, *Night Comes to the Cumberlands,* and Lyndon B. Johnson tried to deploy the power of the federal government to eradicate it.

As they were when I was a public school student in Harrison County, school districts are constantly short of funds even though coal firms are supposed to pay severance taxes for the coal they mine. West Virginia coal firms paid $379 million in severance taxes in 2008—not all that much, considering that the state's tax revenues run about $4 billion annually. Funding is anemic thanks to years of legislative tax breaks to coal firms and diminished real estate assessments caused by surface mining. The pin-money nature of the severance tax was underlined when the West Virginia governor, Earl Ray Tomblin, got into hot water by proposing giving $100 million—or roughly a quarter of the coal firms' annual contribution—to state gambling casinos.

Even if better-educated students do graduate from high school, there are few attractive and lucrative jobs available locally. Then and now, out-migration of talented youth is the norm. According to old saws in Appalachia, you learned "readin', writin', and the road to Baltimore [or Pittsburgh, Cincinnati, Knoxville, take your pick]." The Appalachian diaspora ended in any number of destinations. In the 1950s and 1960s, poor mountain folk flocked to Chicago's Uptown neighborhood. During World War II, whole families and their neighbors in hollows moved to Akron to work in rubber plants, just as they flocked to the shipyards of Norfolk and Newport News. One of the more famous émigrées was the late Hazel Dickens, who moved in the 1950s with her coal-mining family from Mercer County in southern West Virginia to Baltimore, where she became a factory worker and one of the country's best-known experts of mountain folk music, especially coal protest songs. Before that, Washington was a favorite spot. When J. Edgar Hoover was expanding his FBI empire in the 1930s, he needed hundreds of trustworthy female clerks. The bureau recruited heavily from the Appalachians, where young white women were extremely hardworking and reliable. Having been raised in fundamentalist Protestant churches, they took biblical writings literally and proved to be very security conscious and respectful of authority.

The military is another big recipient of Appalachian youth.

Many have Celtic blood and come from a tradition that re-veres the warrior who is quick to fight. Physical toughness is admired. For example, Tommy Davis, the Upper Big Branch miner who lost his son, brother, and nephew in the 2010 blast, says that when his boys were young, they had to exercise on a pull-up bar he had installed in the doorway to the kitchen. "If they went in to get a glass of milk, they had to do ten chin-ups," he told me. Later, when his son was a star wrestler in high school, Tommy Davis loved to attend matches and cheer him on. "I told him, he could go fuck up some city boys. He could fuck 'em up on the wrestling mat or he could fuck 'em up later in the parking lot. Either way."

The tough-guy mystique carries over to coal jobs. Tommy Davis admires how his brother Timmy, who was killed in the mine explosion, handled the physical dangers on the job. "One time, Timmy was in a mine and a roof bolt fell on him and knocked him silly. Had some of his teeth bashed out. I saw him in the hospital and even though he was in great pain, he let me know he could take it."

Attitudes like this are one reason why a disproportion-ately large number of mountain boys ended up in combat zones such as Iraq or Afghanistan and, before that, Vietnam. As a former marine veteran of the war and a U.S. senator, James Webb writes in his 2004 book *Born Fighting: How the Scots-Irish Changed America,* while it was assumed that Viet-nam was fought by the poor and draftees, "More accurately, it was a war fought mainly by volunteers, including two-thirds of those who served and 73 percent of those who died."

Prowess with firearms is another part of the culture. Before going into military service, a number of mountain boys were already expert shots, since gun ownership and hunting are accepted, common practices. Boys got their first guns when they were very young, sometimes under ten years old. My parents were more conservative, so I didn't get my first and only rifle until I was eleven—a Savage 63 bolt-action .22-caliber gun with a full-length Mannlicher hardwood stock. A 2006 survey showed that 55 percent of West Virginians said they had a gun in their home; in Kentucky, the number was 47.7 percent. These are among the highest percentages reported in the United States, with only such hunting-minded states as Wyoming, South Dakota, and Alaska edging them out.

Unfortunately, these and other traits of Appalachian folk have been twisted and expanded for at least two centuries. They have been depicted as fiercely independent, violent, and backward people who don't like change, don't care how they dress, and couldn't care less what others think of them. According to Anthony Harkins, a professor of history at Western Kentucky University, exploiting the buffoon image of the shiftless moonshiner hillbilly and his fecund, pipe-smoking woman traces back to the eighteenth century. That was when the region, which had been left as a hunting ground by Native American tribes, started to draw Scots-Irish after Queen Anne issued her decree in 1704 that elected officials in her realm had to take Anglican sacraments. The crackdown against Presbyterianism and gloomy Calvinism forced many Celts, particularly Ulstermen, to the American colonies. They came through Philadelphia and moved on to the mountains of Pennsylvania and Maryland and then south into Virginia and the Carolinas. The soil was rocky and poor, so the landed plantation gentry of the Tidewater areas weren't interested. Of English descent, that gentry looked down on the Celts and also on the Germans who later settled in the mountains, but the elites did see a way that the tough Scots-Irish might be useful by providing a buffer zone against Native Americans to the west and against Catholics in the Maryland colony. The new white mountaineers started two centuries of isolation.

The tension between the slaveholding Virginians and the mountain people spilled over during the Civil War, when West Virginia seceded from the Old Dominion. Unlike Virginia, West Virginia was not ravaged by the war, although it did see some battles. It muddled along with timber and subsistence farming until, in 1853, David T. Ansted, a British geologist, surveyed an area north of Beckley and east of Charleston. He found great reserves of high-grade bituminous coal. The reserves covered most of the western slopes of the Appalachians from Pennsylvania to Alabama, with the best reserves in West Virginia and Kentucky. As America grew and demanded raw materials for its industrial revolution, the die was set.

Squads of land- and mineral-rights buyers descended upon the isolated mountain families, taking advantage of their food, drink, and naïve hospitality. They offered one-sided contracts that gave them control over land the mountaineers may have held title to since before the Revolutionary War. As Harry Caudill wrote: "Across the table on a puncheon bench sat a man and a woman out of a different age. Still remarkably close to the frontier of a century before, neither of them possessed more than the rudiments of an education. Hardly more than 25 percent of such mineral deeds were signed by grantors who could as much as sign their names."

Caudill's sympathetic and honest image has been distorted over the years for amusement by such cartoonists as Paul Webb and Al Capp, whose *Li'l Abner* with its Dogpatch and Daisy Mae cemented the idea of the uneducated hillbilly rube. Billy DeBeck's *Barney Google and Snuffy Smith* comic strip took offensiveness to new heights in the 1930s as he played off a lazy, drunken hillbilly against high-society types and has him beating up an African American doorman who won't let him in a haughty private club. Walt Disney, hardly a friend to minorities, got into the act with his 1946 *Make Mine Music* animated work featuring "the Martins and the

Coys," who passed out drunk from moonshine when they weren't feuding. A more modern and enduring image of mountain people comes from James Dickey's novel *Deliverance,* which was presented as a movie in 1972. Four suburban Southerners take a quixotic canoe trip in North Georgia that turns into a nightmare of injuries, murder, and homosexual rape so powerful that it became an Appalachian touchstone for many Americans. It was also sheer cultural exploitation. In one scene, actor Ronny Cox plays a remarkable guitar and banjo duet with a cross-eyed boy, who in real life was recruited from a local special education class in Rabun County, Georgia, where the movie was filmed. The boy had his head shaved and was dusted with white powder to make him seem like an inbred albino. Even the rape scene was phony. In the real-life background for the incident, author James Dickey wrecked his canoe but was actually helped out by mountain men after they determined he wasn't a revenue agent looking for moonshine stills. The negative image, however, prevails. Film scholar Pat Arnow wrote in his 1991 article "Hilbilly": "*Deliverance* is still the greatest incentive for many non-Southerners to stay on the Interstate."

Playing the mountain people as childlike, ignorant, and hopeless was far more than a Hollywood machination. For the people who owned coal mines in Central Appalachia, it was a key part of their plan for social order. The scheme was a familiar one in the South of the late nineteenth and early twentieth centuries—to keep the working classes weak, dependent, and unable to create their own middle class. According to Cynthia Duncan, a sociology professor at the University of New Hampshire and director of its Carsey Institute, which studies the region's families and communities, coalfield folk were cemented into the "have-not" equation. She researched coal communities in Eastern Kentucky in the 1980s with their history of stores owned by coal companies that required "script" or a private currency issued by the coal firm.

Taking a cue from cotton plantation bosses in the Mississippi Delta, coal barons deliberately kept a middle class from evolving among the miners. A clear class division was essential for control. "In the Mississippi Delta and in the coalfields it was in the interest of the bosses to keep the workforce vulnerable and dependent," she notes. Her research into a northern New England lumber town that did have a middle class showed a different attitude. In the cotton and coalfields, everyone knew which families controlled everything if they were asked. "In northern New England, the people wouldn't know what you were talking about," she says.

Consequently, the idea of organized labor was threatening on many levels. Its immediate danger was that it scrambled business plans, ruined competitive cost advantages, and jeopardized profits. On a deeper level, it threatened the entire social arrangement that the coal executives had set up and badly wanted to keep. While the years have brought obvious changes, the mentality still persists today. It was employed by Don Blankenship at Massey Energy, where employees were kept off balance by constant purges and then bought off by company-sponsored excursions to Nashville or a water park near Williamsburg, Virginia.

Yet the constant threat of injuries and death from coal mines would not go away, giving the miners ideas about organizing and the bosses new reasons to fret. Body counts would rise into the hundreds. In one example, on December 6, 1907, in a scenario that foreshadowed the tragedy at Upper Big Branch, a methane fire touched off coal dust at the Monongah Mine in Marion County, West Virginia. The official death toll was 362, including 171 Italian immigrants, but some estimates went as high as 500 dead. Another blast on April 28, 1914, at the Eccles Mine west of Beckley killed 186. In yet another display of how big-name interests always seem to have a hand in Appalachian tragedies, the Eccles Mine was owned by the famed Guggenheim family, better known for its philanthropy that funded such institutions as the Guggenheim art museum in New York City.

By the end of the nineteenth century, labor organizing had become common in the industrialized Northeast and Midwest and had been fueled by a flood of new European immigrants who were more willing to buck authority. Their spirit drifted south to the Appalachians, where miners starting thinking about unions. One of the most famous flashpoints in the coming fight over unionizing Central Appalachian coalfields was in the town of Matewan, the town where Don Blankenship was raised. The Matewan Massacre took place on May 20, 1920, culminating after rising tensions between miners and labor activists and the Stone Mountain Coal Company, a nonunion operation that was trying to keep out the union organizing that had swept Northern Appalachian coalfields. The coal firm sent in strikebreakers from the notorious Baldwin-Felts agency. They confronted Sidney Hatfield, the Matewan police chief, leading to a gunfight in the middle of town that killed four men, including the town mayor. The shoot-out immediately made Hatfield a celebrity among the union community, and he was later acquitted of murder. Retribution was coming, however. In broad daylight on August 1, 1921, Hatfield was assassinated on the steps of the McDowell County Courthouse by Baldwin-Felts agents who were never arrested. By some accounts, Hatfield was shot seventeen times.

Back in 1921, the massacre and assassination of Hatfield became a miners' battle cry. In Hatfield's name, they grabbed rifles and started patrols in nearby Logan County. When local law enforcement confronted them, the police were turned back. Militancy grew statewide with calls to march across Blair Mountain to reach Mingo County, where a number of miners had been jailed in a previous altercation. Tempers ran so hot that even Mother Jones urged them to cool off. Armed miners gathered in Marmet, a coal depot on the Kanawha River to the north, and started their march south, gathering miners along the way and commandeering a Chesapeake and Ohio freight train. Their ranks grew to more than thirteen thousand.

The stage was thus set for one of the worst labor-related bloodlettings in U.S. history, although not a word of it ever came up in my classroom back when I learned how to spell "rhododendron." Blair Mountain had been turned into a fortress by the Logan County sheriff, Don Chafin, a coal-industry stooge who recruited a private army of about two thousand. They confronted miners who marched upon them. On August 29, a pitched battle ensued. Aircraft bombed the miners' positions with homemade explosives and gas devices left over from World War I. President Warren Harding ordered in federal troops and army bombers. More than one hundred deaths had been reported before the troops arrived on September 2, ending the battle. Eventually, 985 miners were indicted for murder, treason, and assorted crimes. Some were jailed, but all were paroled in 1925 by state officials seeking mass amnesty for labor peace. While management won the battle, the United Mine Workers won the war, at least until A. T. Massey Coal came into power and took the union apart in the Central Appalachian coalfields sixty years later.

•   •   •

Coal's importance to the mountain workforce, however, has been in marked decline since the 1940s. Coal employment in Central Appalachia peaked at about 475,000 jobs during that decade. Today, only thirty-eight thousand make a living in coal mining, and throughout the region, only 2 percent of direct employment is related to mining. Such data are all the more remarkable, since the circle of disaster, legislative reform, and disaster again remains as stubbornly unbroken as it was a century ago as evidenced by the Upper Big Branch explosion.

Also unbroken is the contemptible idea that common people need to be kept in their place. That was the clear, underlying message on April 5, 2011, when the families of the dead miners killed at Massey Energy's Upper Big Branch were brought in as props for a memorial service. It had been billed as a "private" memorial service for the families of the dead. Yet the ceremony was anything but private. The auditorium of the grade school in Whitesville, West Virginia, was filled with uniformed police officers, reporters, camera crews, and well-scrubbed political aides. Outside, the football field had a scoreboard that read "Us" and "Y'All." Inside onstage were twenty-nine white miners' helmets on crosses, each one holding alternating black and white bows.

Media people crowded in and there were representatives from the state's larger newspapers, the Associated Press, and National Public Radio, which had been covering the disaster doggedly. As if in a grade school assembly, families started entering from the right side of the room. They were young to middle-aged, with some elderly people. Many were clad in blue jeans and had T-shirts commemorating loved ones. IN MEMORY OF GRIFF was printed on one of them. Girls and men wore Massey work shirts, and more than a few had on Harley-Davidson caps or shirts.

The strange service was much more of a political event being staged for the benefit of West Virginia's acting governor, two U.S. senators, three congressmen, and Barack Obama's secretary of labor than for the twenty-nine dead miners and their survivors. Another service at a Beckley church a few hours earlier had been a low-key and reflective event that did, in fact, seem aimed at remembering the dead. But here in Whitesville, exploiting mountain folk continued as it has for decades.

"Let us stand and respect our honored guests," said the announcer. The "honored guests" were not the families of the deceased but the politicians who filed onstage. For the next 110 minutes or so, the Whitesville audience heard from all the politicians, a Protestant minister, a local fire chief, and the West Virginia State Police chaplain, whose human and remorseful talk stood in marked contrast to many of the others. The notes sounded were all the same. The miners were brave. West Virginians should be proud. Coal mining is tough. Acting Governor Earl Ray Tomblin, who took over when Governor Joe Manchin went to Washington to replace the late Senator Robert C. Byrd, said, "Coal mining is a brotherhood.

The reality of what happened is unbelievable." It very well could be, because he said the miners died to "keep the lights on for us all," a sentiment that has been echoed for years and hasn't been true for almost as many. A former member of the state senate, Tomblin is one of the few Democrats who received campaign contributions from Blankenship. U.S. Senator Jay Rockefeller, the patrician-looking great-grandson of oil magnate John D. Rockefeller, described coal mining as a tough job, but the Upper Big Branch men bucked chances to leave their jobs and the state. "I am very proud to look out at you," said Rockefeller, who came to West Virginia in the 1960s as a social worker with the federal Volunteers in Service to America (VISTA) program and liked the state so much that he ended up staying. U.S. Representative Nick Joe Rahall II, first elected to Congress in 1976, sounded similar themes: "We must do all we can to ensure that this never happens again." Despite Rahall's posturing on Upper Big Branch, his political campaigns have been bankrolled by the coal industry for the past thirty-five years. He has received money from Peabody Energy; Patriot Coal; the two coal-hauling railroads in the region, Norfolk Southern and CSX Transportation; as well as the Mountaineer PAC, whose membership includes Alpha Natural Resources. As the politicians droned on, a mother cradled her screaming infant in her arms and walked outside so as not to disturb others.

It finally came to an end when Matt Jones, a local man who wore a Massey shirt and a straw cowboy hat, sang the

John Denver anthem, "Take Me Home, Country Roads." In the middle of the song, Jones choked up. He apologized and said he was the brother-in-law of one of the dead miners.

Diversifying the regional economy away from its dependence upon coal has been a problem for decades. Economic development officials keep trying. They have to leverage their workforce's education and skill levels against other competing rural areas. A popular choice is setting up call centers where telephone orders can be taken for goods or services with minimal training. A problem is that call centers tend to be short-lived, since they are just as easy to shut down in difficult economic times as they are to set up. One exception appears to be Logisticare, an Atlanta-based firm that offers non-emergency transportation for sick people. In 2002, it used public funding to renovate an old railroad hotel in the coal town of Norton, Virginia, into a call center that employs up to a hundred people.

There also have been efforts to upgrade neglected education systems. In the coal towns of Harlan and Hazard, the Kentucky Educational Reform Act (KERA) program during the 1990s improved grade school performance statewide. West Virginia and Virginia have likewise begun statewide competency testing in public schools, and higher learning centers such as Marshall University and the University of

Virginia at Wise offer many distance and community-outreach educational programs.

Not all the stories are uplifting. In early 2011, the state-wide unemployment rate in the Mountain State was above 9 percent, and most counties in the richest coal-seam counties reported rates of higher than 10 percent. Outside of jobs in coal mining or call centers the only employment is in retail at stores such as Walmarts in larger towns. Antiunion and tight on benefits, Walmart has its own bundle of employment issues.

While the range of jobs is limited, there's actually a shortage of experienced miners. This inconvenient truth flies in the face of otherwise high levels of unemployment and shows, instead, just how unpopular and risky coal jobs are considered to be. Indeed, on April 5, 2011, just after a memorial service at a Beckley church ended for the twenty-nine dead miners at Massey Energy's Upper Big Branch Mine, a radio advertisement hawked mining jobs for Massey. A jobs fair was to be held at Chapmanville, about thirty-five miles southwest of Charleston. Massey was offering full benefits, good pay, and a chance to be part of the energy future. They have to, because in spite of the lack of such employment, the dangers of mining still have workers looking elsewhere for employment.

While employers praise mountain folk for their hard work and reliability, another nettlesome issue is emerging.

Appalachia has higher-than-average levels of drug abuse and mental illness, according to the Appalachian Regional Commission. It reports that more people living in Appalachia have psychological distress or major depression—13.5 percent of all adults regionally compared to 11.6 percent nationally. One result is that drug abuse has expanded quickly in the mountains, particularly involving methamphetamines and oxycodone, also known as OxyContin, the prescription pain reliever. Known colloquially as "hillbilly heroin," OxyContin is an especially fast-rising scourge. At one major Southern university, a student from the Southwest Virginia coalfields displayed her high school yearbook and easily pointed out who the OxyContin users were. Since OxyContin reduces calcium and dries out their mouths, they were the ones who had bad teeth, gums, and other obvious dental problems.

The drug issue impacts employment, exacerbating the contradictions in the labor situation. In a region where there are few jobs, high-paying slots often go unfilled because companies can't find people who are qualified, willing, and can pass pre-employment drug screening. Norfolk Southern, for example, scrambled to hire workers in anticipation of the metallurgical coal boom that kicked up steam in 2010. A big problem, Mark Bower—who heads traffic for the railroad—told a conference, is finding qualified workers. "If you hold an employment event at a civic center in Charleston, you'll get one hundred people applying for jobs, which is good. But after you do drug tests and the background checks, you'll have maybe ten. And out of those, you'll only want to make offers to about three." The same pattern exists in hiring for coal mine jobs. Many may apply, but after drug testing and actually seeing what the work conditions are, a majority drop out, leaving coal firms short of people. Some skimped on drug testing, but not Massey Energy, which, seemingly out of character, had a reputation for insisting on pre-employment drug testing and was known to have regular spot checks on the job.

Nothing ever seems easy for the mountain folk of Central Appalachia. Their stubborn independence and warm hospitality are perpetually tempered by conditions and employers far beyond their control. It is almost as if the region has a special curse that becomes evident when one drives west of U.S. 19, which is where most of the coalfields begin. Sweeping vistas with blue-green mountains far in the background give way to a sense of being shut in by steep hills that alter life cycles. Winter twilight comes especially early as the sun's rays are cut off by mountaintops. Summer humidity takes on a more languid quality, since the valleys can be too steep for wind to blow it away.

Everywhere is evidence of the coal that has defined the region. There is the ubiquitous soot, the forlorn dirt roads with the occasional faded memorial to dead miners, and the railroad spur lines that follow creek beds and whose endless streams of hopper cars clatter slowly by and keep motorists trapped at crossings seemingly forever.

Local folk seem still trapped as well by the coal culture that has long exploited them while offering few other options. Although there are far fewer coal jobs today, people remain frozen in a class system that was defined years ago by coal, with the boss man in the house on the hill and more ordinary mortals in the double-wides or leftover coal-camp cabins. The coalfields have their outdoor pleasures. Even so, it is clear that while the working people stay, the wealth has gone.

4

THE ROOTS OF MASSEY ENERGY

E. Morgan Massey, the eighty-five-year-old great-grandson of one of Massey Energy's founders, stretches back in his first-floor office in the Richmond firm's dark, dour-looking headquarters building, a black, marble-clad structure that he helped design in 1950. Attired in a gray sports coat, slacks, and a checkered shirt, his pale blue eyes shining brightly, Massey has a Southern charm as he discusses his family's history. Behind his large desk is an abstract work in burnt reds and browns of a mountain landscape painted by a friend at Virginia Tech. On the side are maps of Venezuela and Colombia, where Evans Energy investment firm, which he now owns, has coal operations, plus China, where he has been doing coal business for two decades.

His wife, Joan, ill with Parkinson's disease, sits at one side of the room. E. Morgan dotes on his wife, who has a tote bag lettered SAILFISH POINT, from Stuart, Florida, where they have a home. His other home is a town house on Richmond's Monument Avenue, a grand boulevard modeled vaguely after Paris's Champs-Élysées that is marked by dogwoods, magnolias, and statues of the Confederate generals Robert E. Lee, J. E. B. Stuart, and Thomas "Stonewall" Jackson. He commutes there from Richmond in one of his two corporate jets, a Citation CJ2+ and a Citation CJ1, each worth about $4.5 million. He's still a pilot, having learned to fly during army training in World War II. "I'm a member of the UFOs," he says with a laugh, "the United Flying Octogenarians."

The family may be a relative latecomer by Richmond's snooty standards, but even in the former capital of the Confederacy—where it can be decades before a newcomer is accepted socially—money talks. Coal money pumped out of the West Virginia hills by the Masseys over at least six decades begat the Massey Foundation in 1957, now worth more than $47 million. It supports college scholarships to worthy students in Virginia and West Virginia, jobs-training programs, the arts, and industrial-safety and medical research. The largesse extends across downtown Richmond to the VCU Massey Cancer Center, part of Virginia Commonwealth University Health System and formerly the Medical College of Virginia. It is one of sixty-six cancer centers out of fifteen hundred selected by the National Cancer Institute to take a leading role in researching the disease. Its five hundred researchers take promising results from labs to clinical trials and then into hospital use. Another beneficiary is the Virginia

Institute of Marine Science, a group based in Gloucester, Virginia, that researches such aquatic ecosystems as the Chesapeake Bay and the Atlantic Ocean. One foundation contribution of $1 million in 2005 helped VIMS weather state budget cuts. But perhaps the favorite work of E. Morgan Massey, a vice president of the foundation, is his current plan with Virginia Tech to research ways to detect explosive levels of coal dust in mines and shut them down automatically.

A. T. Massey Coal Company, Inc., the predecessor firm of Massey Energy, may have had generally good relations with its employees and was generous with philanthropy, but it had an iron fist when it came to unions. The hard lessons it learned battling for a profit margin in the fickle coalfields made it a stickler for watching expenses. That said, A. T. Massey also had a reputation for letting its workers have their say—unlike what Massey Energy became under hard-charging Donald Blankenship.

And E. Morgan Massey is skeptical of the merger, noting sarcastically, "Alpha is union and Massey is nonunion. It's going to make for a nice combination."

The origin of the Massey family in the coal business is not in Richmond but 250 miles to the west in Powellton, an unincorporated flyspeck of a town near Charleston, West Virginia. Today, it is a collection of small houses along Armstrong Creek near Black Gunn Hollow with a grocery named Angela's as its only business. This tiny burg was where William Evans, a sheep-herder and E. Morgan Massey's great-grandfather, emigrated from the town of Llanidloes in the county of Powys, Wales, in 1876 to go into the coal-mining business, in which he had some experience from the Welsh pits. With him was his Welsh partner Evan Powell, for whom Powellton was later named.

At the time, the state of West Virginia was only thirteen years old, its voters having seceded from Virginia in 1863 following disputes over slavery, landownership, and service in Confederate forces during the Civil War. Moving there was typical for a Welshman who faced much of the same scorn from those of English ancestry who had settled in the Tidewater, Virginia, area as the Celts who preceded them. A big change was pending, however, as the Chesapeake and Ohio Railway had just completed a line from the New River Gorge in West Virginia through the Allegheny Mountains to ports around James River and Chesapeake Bay, which gave them the opportunity to ship coal to Philadelphia and points north, including the metallurgical coal that went to steel mills.

What started as Powell Coal changed names to Armstrong Coal. The company needed buildings to operate, especially a coal tipple to load coal onto railcars. Helping was a skilled carpenter in the area named Antonio T. Massey, a tobacco farmer from Stuart, Virginia, who had moved to the Mountain State with his brothers Robert and James. Seeing an opportunity, they started their own coal business in the small town of Ansted, not far from Powellton. The firm, Mill

Creek Colliery, produced and sold coal to heat apartments in Chicago and also to electric utilities and railroads.

The new coal operators struggled in the primitive conditions. At the time, mining was a rudimentary affair with little thought given to air ventilation or other safety measures. Miners got their light from candles fixed on their helmets. Highly flammable methane gas tended to accumulate at the roofs of the mines, since it was lighter than air. With the proper mix of oxygen, it would explode; otherwise, it would just burn. In order to work in the shafts, miners had to routinely burn off methane, and in an eerie foreshadowing of the Upper Big Branch disaster, a controlled methane burn went out of control and William Evans was injured so seriously that he died a few weeks later. The mine-shaft floor where toxic carbon monoxide (also known as "blackdamp") collected was no less dangerous. Caged canaries were brought in and set on the floor. If they suffocated, it was a warning that the carbon monoxide mix was potentially deadly. With the methane above and the carbon monoxide below, miners were constantly caught between two potentially lethal dangers.

A. T. Massey spent his time on the surface after ending his mining career in 1897. He had special advantage. In 1894, he married Sarah Jane Evans, William Evans's daughter, who had eventually followed her father from Wales to West Virginia. When she first immigrated to the wilds of West Virginia from Wales, she wrote, "I had moved to the ends of the earth." Their wedding took place at the Grand Hotel in

Thurmond, West Virginia, a fast-growing railroad town in the New River Gorge that would soon become infamous for gambling and prostitution. Miners howled at the moon and shot out the lights after they drew their pay. The gorge at the New River, the second oldest river in the world, had practically overnight become a major center of the booming coal industry. Shaft mines were bored into the high sides of the gorge towering a thousand feet up. Narrow-gauge trains, mule teams, or crude, wooden conveyor systems carried the product to the riverside, the location of the main line of the Chesapeake and Ohio Railway, which hauled the coal east to Tidewater ports or west to the burgeoning Midwestern cities of Chicago, Cincinnati, Detroit, and Cleveland as well as the blast furnaces in Pittsburgh, Baltimore, and Birmingham. New River helped supply them. Filling the gorge with thick, putrid-smelling smoke, scores of coke ovens—brick contraptions shaped like oversized beehives—worked night and day. Completing the vast ecological destruction, mining firms hacked down thousands of acres of virgin forest to make way for King Coal.

Thurmond, situated almost in the middle of the gorge, was the focus of the fast money. At a hotel known as the Dun Glen, what was billed as the world's longest-running poker game was supposed to have taken place. The town's Balahack neighborhood became a notorious red-light district and perhaps a reason why another local hotel, the Lafayette, was known colloquially as the "Lay-flat."

The Masseys were blessed with resources beyond the grasp of most coal-town folk. Sarah had deeds to land and coal reserves given to her by her father. Still, when triplets followed the birth of her first son, Ivor, the country doctor was drunk, so they had to put him to bed and send for another doctor fifty miles away.

A.T., who had been a deputy sheriff in Raleigh County, took a job in Glen Jean, near Beckley, running a coal company store that provided goods to miners in exchange for script. A fire destroyed the store in 1910, sending A.T. to the New River Coal Company, where he was a broker. In 1916, the new job took A.T. to Richmond, the largest city on the east side of the C&O main line, which handled trainloads of coal from the Central Appalachian fields to Virginia seaports and markets farther away. Richmond was by then a bustling city with many attractions that Glen Jean lacked, such as fancy restaurants, museums, and several large, well-appointed department stores that drew shoppers from the outskirts of Northern Virginia to the tobacco fields of eastern North Carolina. For Ivor, the triplets, and Evan, it was to be a richer, fuller life.

A.T. had a knack for selling coal. World War I had boosted demand for steel and coking coal as the United States ramped up production of rifles, helmets, and ships. He disposed of some mine interests while marketing coal from the Evans family mines under the new name of A. T. Massey Coal Company, Inc., in 1920. As his wealth grew, so did A.T.'s venal tendencies. "He drove as fast as he could and was very athletic and good-looking. He had a lot of girlfriends and gave out ten-dollar bills to poor people. A real salesman type," states his grandson. "He was a rounder."

The firm by now had offices in Cincinnati, Chicago, and Cleveland to market its product. But A.T. was also using them as entertainment centers. His free-spending ways were threatening the family nest egg. In time, Evan, his son and E. Morgan's father, had moved up in company management. On July 27, 1934, he wrote his father a letter in care of the posh, beaux arts Sinton Hotel in Cincinnati scolding him for racking up expenses for June 1934 in the amount of $1,480. Evan reminded his father that he had agreed to spend only $500 a month, but had been, in fact, spending $1,500 a month for the past several months. "Now no small company can have this heavy expense as your salary and expenses alone for you and your family are amounting to at least one fourth of the gross income of the company. This is absolutely ridiculous and has to be stopped." Evan ended his missive: "This letter is written with the best intentions and the best wishes for you and your future."

A.T. however, continued to drain the firm and his family's wealth. As he spent money, his family lived modestly in a Victorian-style house on a sprawling green yard at 1203 Wilmington Avenue, in Richmond's quaint Ginter Park neighborhood. Today, the house is painted pink with an intricate white wooden lattice railing around its ample front porch. In 1945, when A.T. died, the firm was $250,000 in the hole, loaded with liabilities, and with no assets "other than the office furniture," says E. Morgan. "Nothing was being left to the seven children." Most of the debt was owed to State-Planters Bank, secured on accounts receivable. Evan and his brother William guaranteed the debt on a 60 to 40 basis and took a like proportion in worthless stock.

Their aim was not just to salvage the company but to grow it as well. With World War II, A. T. Massey Coal would change again. Coal sales were skyrocketing, and the firm also started to get involved in coal production, not just sales. With Evan and William E. at the helm after A.T.'s death in 1945, the firm started buying up other coal firms, such as Gay Coal & Coke and, later, the Royalty Smokeless Coal Company, which mined the Fire Creek coal seam next to Babcock State Park in Fayette County. New reserves in Mingo and Logan counties were added to the inventory as Massey began to experiment with Joy Mining Machinery, a major maker of coal mining equipment, to come up with special gear to handle the peculiarities of Massey's new mines. A crown jewel was the purchase of Omar Mining Company, owner of the largest coal mine in southern West Virginia. And E. Morgan, then an army veteran and recent college graduate with plenty of hands-on experience in family coal mines, was pushing for bigger and more acquisitions.

Another postwar venture that would prove enormously valuable came through an unlikely source: Hugo Stinnes, a German industrialist who had acted as an adviser to Adolf

Hitler and his war machine. In 1924, funding from the Stinnes family helped the Nazi Party ramp up publishing a news-paper from a weekly to a daily, and he was one of a group of prominent German industrialists, including arms maker Krupp and steel man Fritz Thyssen, who had bankrolled the Nazis. Later Stinnes became disillusioned with the Führer's ruthlessness. A war crimes tribunal exonerated him after Ger-many surrendered. Immigrating to the United States, Stinnes got into business with Morgan Massey's father and became a big influence on the coal brokerage.

Most of the coal mines in Europe had been destroyed, "so A. T. Massey started shipping coal to France and Germany in ten-thousand-ton Liberty ships. It created a boom in the coal business in 1946–47," E. Morgan says. The export model was cemented when Massey upgraded to new, twenty-five-thousand-ton ships for the postwar Japanese trade. Over the next six decades, foreign exports would become a profitable asset, helping A. T. Massey overcome decreased demand for coal as trains shifted to diesel engines and travel shifted to gasoline-powered automobiles using the new interstate sys-tem and commercial aircraft. Longer term, the move would prove even more prescient. In the 1950s and 1960s, Japan and Germany would rebuild their economies to become global economic powerhouses. Later still, in the 1980s and 1990s, China, India, and Brazil would grow beyond expectations. By 2009, despite a devastating recession in the United States, China and India would continue their expansion, which called for lots more cars, buildings, bridges, and highways for two billion people. All would need coal and lots of it, and Massey had what they needed.

E. Morgan Massey, the last family member to head the firm, matured under the protective wings of his father, Evan, who recognized that the youngster had a knack for things mechanical. E. Morgan said, "He let me overhaul a company Buick when I was only fourteen. It worked fine, but after a while the rings I had worked on started knocking. I didn't know about the torquing that you were supposed to do with the rings. He sold the car real quick." At seventeen, E. Morgan started working as a "field man" in the coalfields and made key contacts that would help him later when he was in charge and growing the firm.

After studying engineering at the University of Virginia for two years, he was called up for service in the Army Air Corps and trained as a pilot, but with the war ending, he was released before he saw action. It was back to the coalfields until he returned to Charlottesville and the university to finish his engineering degree.

His father, Evan, was running the company along with his uncle, but E. Morgan kept pushing to get heavily into production by recommending that they buy up key mines, including the 1954 purchase of the West Virginia Coal and Coke Omar mine that played a substantial role in exporting coal to war-torn Germany. Massey paid $250,000 for the property plus royalties on coal sales. At times, E. Morgan didn't tell his dad of his purchases. According to a 1985 *BusinessWeek* profile, his first acquisition, "a $10,000 investment in a coal venture was [made] without his father's knowledge."

As always, the coal market had its ups and downs. By 1952, the coal business was in a recession, but E. Morgan was able to bottom-fish for distressed coal properties that his father could pick up for a song. One such tract was in the rich Sewell coal seam near the New River Gorge. The mine had big problems: Its seams were thin, and most were high above the New River railroad main line and reachable only by an antiquated narrow-gauge train. He picked up thirteen thousand acres for about two dollars an acre but was able to make big profits later, after rail lines and coal markets improved. In another case, he bought a distressed property and "put contract miners in the mines" while waiting for the market to turn. When it did, in 1970, the mine became extremely profitable.

Young E. Morgan was also spending a lot of time in the mines, which was unusual for a company heir apparent. He labored as a miner at Royalty Smokeless Coal with ten men loading four hundred tons a day in ten-hour shifts. Tapping his engineering skills, he tinkered with new forms of continuous mining machines and schemes for conveyor belts to haul coal to the surface. Working in such a way had a personal price for E. Morgan. When his first wife became pregnant, Massey says, she didn't want to risk being in an area isolated from medical care, so he had to commute to and from Richmond.

For the first time, A. T. Massey Coal also ventured into surface or strip-mining, a practice that was becoming widespread in the 1950s Appalachians as surplus diesel-powered earthmoving equipment left over after the war became available. Until a 1977 federal law, the practice left hills from Pennsylvania to Alabama with ugly brown gashes on their sides, and uncontrolled runoff from coal waste polluted streams. The impacts of this form of strip-mining, however, cannot compare with the effects of today's mountaintop-removal practices, which can involve hundreds if not thousands of acres at a single mine.

Hands-on experience also affected E. Morgan as he developed his managerial style. Lessons learned included keeping a sharp eye on production costs and avoiding unions wherever possible. In later years, he summed up his personal mission statement the following way: Customers came first, followed by shareholders. Employees rank third on his list of concerns, followed by the community and the environment. "Nothing happens without money from sales from customers," he says. Shareholders are in second place because they can be intrusive interlopers unfamiliar with the real problems of the business. Employees are there to work, but can't unless the money is flowing and the needs of shareholders are covered. So, by Massey's credo, community and the environment barely make the cut.

The hard-charging young manager soon got into scrapes with unions. In what would be a trademark Massey move, he changed the company that ran the West Virginia Coal and Coke property at Omar. In doing so, he cut what had been a 2,400-employee workforce to 1,800 people and won union animosity. The company would buy up mines. If they were unionized, management would suspend production until the labor contracts ran out. Then the mine would be reopened as a nonunion mine. The modus operandi became known as the Massey Doctrine, and Blankenship later kept up the practice. When Massey bought the unionized Cannelton Mine in West Virginia after its parent went bankrupt in 2004, Blankenship hired a few union workers but kept most of the workforce unorganized. Another common scheme was to divvy up the work at new properties. The Massey firm would work the best and easiest-to-mine coal. The rest would be farmed out to contract mining firms that often had weaker safety standards, were nonunion, and had much higher management turnover rates, leading to a higher percentage of accidents.

Morgan turned to subterfuge to hide his planning from union bosses. In one case, a new coal property was named Elm Development Company, with the letters in "Elm" being the first letters of the last names of Massey managers. "We didn't name it Elm Coal," he says. "That way the UMWA missed it and we could get by with being non-union."

The firm also had gotten into the practice of going down-market in buying coal-mining equipment, a move that some of the firm's managers opposed. One official was transferred to a new job when he insisted that the company buy top-of-the-line Joy mining gear. "We always bought secondhand equipment and rebuilt it ourselves, and we used contract mining," Morgan says. This hand-me-down mentality became part of the company's culture and would come back to haunt Massey Energy when the use of faulty, aging equipment triggered the deadly disaster at Upper Big Branch.

As its bank accounts grew, A. T. Massey Coal set up the Massey Foundation in 1957, which paid the way for the family's acceptance in Richmond high society. In the 1950s, the firm bought the first of its corporate aircraft that let executives avoid the torturous, time-consuming car trips up and down the dogleg roads of the West Virginia and Kentucky mountains. As the acquisitions piled up, Massey's revenues skyrocketed, eventually soaring from $100 million in the late 1970s to nearly $1 billion by the late 1990s.

Although it couldn't have been known at the time, the practice of buying cheap gear inadvertently pushed A. T. Massey much further than its founders could ever have imagined. As was so often the case in the Appalachian coalfields, another tragic mine disaster had brought along another round of new regulations that required coal operators to change how they did business. And, as the Upper Big Branch fatalities showed years later, the regulations were never enough to prevent the horrors from happening again.

At five thirty on the morning of November 20, 1968, ninety-nine miners were working in a mine owned by Consolidation Coal a few miles west of Fairmont in northern West Virginia. A blast at the mine shook the surrounding land for twelve miles, and flames shot 150 feet above mine shaft openings. Within a few hours, twenty-one miners made it to the surface, but seventy-eight remained trapped underground. Fire kept rescuers away for a week, and, certain there were no survivors, Consol sealed the mine ten days later. The bodies of fifty-nine of the dead would not be recovered for another nineteen years, and nineteen remain entombed permanently. An exact cause was never established, but the mine, which first opened in 1911, was known to have had problems with air ventilation, methane gas, and coal dust—again, just like Upper Big Branch forty-two years later.

The following year, in 1969, Congress passed the Federal Coal Mine Health and Safety Act, which created MSHA and increased the number of annual inspections to two for surface mines and four for deep mines. Monetary and criminal penalties were established for serious violations of safety regulations. Compensation for crippling pneumoconiosis, or black lung disease, particularly for older miners, was set up. As with many mine laws, the act did not prevent what it intended. A year later, on the anniversary of the day it was passed, December 30, 1970, there was an explosion at the Hurricane Creek Mine in Hyden, Kentucky, 2,400 feet below the surface. The small, nonunion operation had been cited repeatedly and shut down by federal regulators during the previous year. Rescue operations were hampered by a foot of snow, and thirty-eight miners were killed.

The new law did have a huge impact on A. T. Massey Coal, however. The flood of new rules meant that the company could not continue its cheap ways of buying and upgrading secondhand equipment. To do that, the company needed capital, and lots of it. The only way to locate that amount of capital was to go public and tap the stock markets. So the Masseys maneuvered to become part of St. Joe Minerals Corporation, a global mining concern, while still maintaining operational control of the firm that retained its name. E. Morgan, then forty-six, took over as president in 1972. The purchase was finalized two years later when he, his brother, and an uncle sold the firm for 14 percent of St. Joe stock worth $56 million. Funds were not only available to upgrade gear but also plentiful enough to underwrite a huge expansion of reserves of lower-sulfur and higher-heat coal that would be needed when federal laws cracked down on polluting electric utilities in the 1970s, and yet again, A. T. Massey was poised to grow bigger still.

Coal got an unexpected boost in 1973 when world oil prices skyrocketed after Arab forces attacked Israel in the Yom Kippur War. Another oil shock after the 1979 Iranian

Revolution and seizure of the U.S. Embassy in Tehran spiked world petroleum prices once more. In addition, electric utilities, especially in Europe, had shifted from oil to coal as oil prices soared.

The wave of petrodollars washed over the bank accounts of big oil firms, which began a new round of acquisitions on other types of energy, such as coal, oil shale, and thermal energy. Companies such as Arco and Exxon started developing the Powder River Basin in Wyoming and Montana, where reserves of low-sulfur, subbituminous coal—which could easily meet federal air-pollution restrictions that electric utilities had to follow—lay in fat seams sometimes one hundred feet thick. The landscape rolled gently, making surface mining costs far less than what they were in the up-and-down hollows of the Appalachians. The traditional coalfields were not left out, however. In 1980, St. Joe sold half of A. T. Massey Coal to the Royal Dutch Shell Group, renaming it Massey Coal Partnership. A year later, Los Angeles–based engineering and construction giant Fluor Corporation bought out St. Joe Minerals, including half of Massey. It picked up the other half in 1987 and started treating it like a cash cow that it otherwise ignored.

For E. Morgan Massey, the 1980s were momentous years, especially for two events: One was the hiring of Don Blankenship. The other was E. Morgan's all-out war with the United

THE ROOTS OF MASSEY ENERGY

Mine Workers of America, in which he broke the back of six decades of union influence in the Central Appalachian coalfields.

After two big booms in the 1970s, the coal market had settled into a prolonged recession by the 1980s. As did all operators, A. T. Massey scrambled to find ways to cut costs, since coal prices were flat or depressed. E. Morgan turned to his tried-and-true cost-cutting ways, such as bringing in more nonunion contract miners who might accept less pay and slip by with less expensive mining equipment. At the same time, the UMWA was seeing its heft in the Appalachians grow weaker because big corporations, many of them nonunion oil firms, were busy developing the Powder River Basin. The results were an even weaker UMWA and a major rollback in pay, benefits, and safety standards for miners. Union membership dropped from 120,000 in 1978 to about 14,000 today, and A. T. Massey led the way.

The flash point was the so-called Bituminous Coal Operators Association (BCOA) agreement, which acted as a kind of omnibus, industry-wide labor contract. Negotiated when the union had considerably more power, the agreement standardized benefits across individual coal companies. If one coal operator fudged, the UMWA would threaten to strike the entire industry in an enormous, preemptive attack. It kept coal barons in line for years. When Richard Trumka, now head of the AFL–CIO, became president of the UMWA in December 1983, he abandoned the all-encompassing strike aspects of

99

the BCOA in favor of a strategy of "selective strikes" against errant coal firms.

Strongly anti-union, E. Morgan Massey seized the opening and notified the UMWA that it would no longer negotiate in a company-wide way; the union would now have to hack out individual agreements with fourteen supposedly independent Massey subsidiaries. It was a "divide and conquer" approach that had served A. T. Massey well for years. The union promptly struck Massey operations in Pennsylvania, West Virginia, and Kentucky. After a five-month-long standoff, the dispute erupted into war reminiscent of the "Bloody Mingo" days of the 1920s, with machine guns, bullets, hired paramilitary thugs, and guerrilla attacks. Royal Dutch Shell bankrolled security that included armored personnel carriers and helicopters. State police and private security men protected "scab" workers as they went to and from their mine jobs.

As *Time* magazine's Frank Trippet wrote in August 1985:

> Violence has become almost monotonous. In the latest incident, a midnight explosion last week rocked the three-story brick district headquarters of the U.M.W. in Pikeville, Ky., incidentally shattering the huge portrait of the late union leader John L. Lewis that hung on the wall. The strike had produced one death, hundreds of injuries and more than a thousand episodes of rock-throwing, smashed windshields, and punctured tires.

Gunfire has become commonplace. Snipers killed a non-union coal-truck driver, Hayes West, 35, in a convoy crossing Coeburn Mountain in late May.

This was the same strike where eleven bullets smashed into thirty-five-year-old Don Blankenship's office, ruining his TV and forever hardening and radicalizing his view on unions.

After the UMWA fight ended, E. Morgan started strategizing about the new realities of the coal markets. Air-pollution regulations had placed the coal found in the heart of his operational area in great demand. It typically had a heat value of 13,000 British thermal units per pound and sulfur levels were 0.7 percent, automatically making it "compliance" coal, meaning electric utilities could burn it while spending less on antipollution gear such as scrubbers. Other Appalachian coal, such as that in northern West Virginia and Pennsylvania, put out almost as much heat but was higher in sulfur. Massey also wanted to leverage his metallurgical coal reserves that "could go either way," he says.

Massey, however, was growing increasingly frustrated with Fluor, even though he was on the company's board. He would fly to Los Angeles for regular board meetings and waste his time as other directors droned on about the myriad businesses the global company had. He said, "They gave me something like fifteen minutes, and Massey was providing them with a third of their revenues. They just didn't understand coal."

Blankenship, meanwhile, was rising quickly in the organization. He was named president of A. T. Massey in 1990 and was even considered later as a potential candidate as the head of Fluor, but lost out when doubts arose that he wasn't worldly enough to handle a global corporation. In 1992, with Massey's strong backing, Blankenship became chairman and chief executive officer A. T. Massey Coal Company.

Soon after, Massey retired, and over the decade of the 1990s, he liquidated his stock, though he kept an office for years on the first floor of the Massey headquarters building in Richmond. He then started Evans Energy, which developed international coal projects in Colombia, Venezuela, and China. Colombia, he says, has "some of the prettiest coal I've ever seen." High in heat and low in pollutants, "it is so clean, it doesn't have to be washed," he says. Venezuela also has good coal and a strategic location for global markets. As good as that sounds, the politically conservative Massey could not stand Venezuela's leftist strongman Hugo Chávez and ramped down his operations there. "Chávez only went to military school. He's uneducated and he doesn't understand things. He hiked up tariffs on things like industrial tires, so when the trucks in his coal mines blow tires, he can't get any and the operation shuts down." Massey found Chinese Communists easier to deal with and has a number of ventures in China.

In addition to his global ventures, Massey still keeps a hand in Richmond. In 2005, he started plans to build a house on an extremely narrow lot on Richmond's ceremonial Monument

Avenue, not far from the Lombardy Street abode where he was born in 1926.

He was deeply troubled by the Upper Big Branch disaster, but he insists it was caused by a natural seismic event. He defended his heir, Don Blankenship, right down to the wire, lobbying with key Massey Energy directors and personal allies Stanley Suboleski, James Crawford, and Baxter Phillips. He even sent lead director Bobby Ray Inman a Christmas card in 2010 urging him to keep the company independent and not to force out Blankenship.

It was a battle he lost, showing just how much Morgan Massey has come to rely on wishful thinking to protect his family's name.

5

BIG COAL'S UGLY AND BRIGHT FUTURE

The PGA National Resort and Spa, otherwise known as "the Home of the 2011 Honda Classic," is hidden behind the palm trees and bougainvillea amongst the strip malls with upscale shops in Palm Beach Gardens, just a few miles and a few highway cloverleaves west of Palm Beach, Florida's best-known residence for the ultrarich. The lush golf resort has a lengthy clubhouse with a red tile roof and Mediterranean architecture. Hallways are covered with pictures of PGA champions, and from the club's balconies one can see the man-made sand traps and water hazards along the manicured fairways. Closer in is the curving pool lined with red umbrellas. A little beyond, a man in a green polo shirt and white slacks quietly practices his putting.

It is a natural place for coal-industry executives and their bankers, suppliers, and shippers, who are gathering in a conference room for the start of a semiannual coal market conference on March 15, 2011, put on by Platts, the large trade publication and data-gathering unit. A group of thirty-something coal market analysts set up their laptops and paper pads as a gray-haired woman in a red jacket calls out to a friend about getting together later. "As long as you're not an environmentalist!" she shouts out as a few people around her laugh.

The mood was optimistic. It had been for weeks before the conference. A series of unrelated events was improving the possibilities for coal in the near and middle term. The promising news was that coal, much-maligned for being filthy and fatal, was looking like it was going to remain a highly important energy option for years to come. Sales of steam coal to power electricity plants were healthy albeit slower because of the economic downturn, and demand was red-hot for metallurgical coal in fast-growing Asia.

Politics were looking better, too. The stubborn global recession and the weakness of the administration of President Barack Obama had given hard-right politicians an opening to battle new regulations that would limit greenhouse gas emissions from coal plants that cause weather change. Despite overwhelming scientific data to the contrary, the very idea that humans contributed to carbon dioxide pollution was under attack by a new breed of young, ambitious, and extremely conservative politicians.

The clear downside to the recession was that big banks were still smarting from the 2008 financial meltdown and were loath to lend much money for new and cleaner electricity plants that used coal. Even so, there was another unexpected reason for the attendees to be in a good mood. It would have been bad form to display it too openly at the coal conference in tony Palm Beach Gardens, but just a few days before the conference began, an earthquake and tsunami in Japan on March 11, 2011, had shattered the Fukushima Daiichi Nuclear Power Plant on the island nation's rocky eastern coast. With three reactors completely melting down, it would be the worst nuclear disaster in a quarter century. Since nuclear is coal's biggest competitor for power generation by utilities, the news was good for coal. The real muscles of electricity power stations are ones that are baseloaded, meaning that they are big enough to generate enough power to meet the power needs of a utility's "base" of customers. Such stations are typically coal-, natural gas–, or nuclear-driven and can provide more megawatts than alternative forms, such as wind, solar, or hydroelectric energy, can supply. Over the past two decades, nuclear power seemed to be operating so safely and cleanly that it was getting fresh looks as the way to go for the future. It did not emit great gobs of climate-changing carbon dioxide, didn't require the mass destruction of Appalachian mountaintops or Colombian jungles, and didn't lead to deaths deep inside coal mines. Fukushima Daiichi would change all of that, but it wasn't completely clear yet at the Platts conference in sunny Florida.

Another thing going for the coal executives was the petering out of the antipollution movement to end or severely curtail the construction and operations of coal-fired plants. The Sierra Club, for instance, had a massive campaign called "Beyond Coal" to drop the energy source. New York mayor and media mogul Michael Bloomberg had even put up $50 million to wean the country away from coal. But with recession-related issues, environmentalists and climate-change critics had pretty much reached their apogee with the 2006 publication of Al Gore's book on climate change, *An Inconvenient Truth,* which would help the former vice president be awarded the Nobel Prize. Leaders in the green movement were ecstatic that Barack Obama had won the presidency in 2008. They expected serious new laws and regulations within the first year or two of his administration. Obama concentrated instead on health care, although he did put $350 million or so into his first stimulus package to help back clean coal technologies and investigate alternative sources of energy, but the effort was for naught. The economy failed to revive after his stimulus program took effect, giving a huge opening to his right-wing critics and Tea Partyers, who made budget cutting and antiregulation the new litmus tests for governing. Clean coal tech and laws to cut greenhouses gases seemed to be going nowhere. "When Obama was elected, we felt pretty confident that he'd be able to enact greenhouse gas legislation," says David Hawkins, director of the climate program at the Natural Resources Defense Council. "At this point, we don't have the basis for change, but it will happen in a year and a half, or two and a half years, or four years," he says.

On other fronts, the U.S. Environmental Protection Agency, moribund after eight years of George W. Bush, would take some new steps to police ruinous mountaintop-removal strip-mining and stem air pollution at coal-burning power plants that could close some of the older ones. Nevertheless, coal remains on a roll. Mountaintop removal is still a viable mining option, and some of the older plants are huge polluters. Overall, 10 percent of the older units produce 40 percent of the carbon dioxide emissions in the United States. They are old, dating back to the 1950s and 1960s, inefficient, and some were earmarked for closing anyway. Coal executives, however, see the closings as part of a "perfect storm" of Obama-led regulation that will stymie their business. One of them is Kevin S. Crutchfield, the chairman and chief executive of Alpha Natural Resources, a firm founded in 2002 and expanded by buying up older, fallen coal companies such as Pittston and now Massey Energy. Dressed in the conservative but sophisticated corporate casual uniform of a blue sweater vest and khaki trousers, Crutchfield later spoke to me at Alpha's headquarters in Southwest Virginia. That day, his talk was peppered with market realism, notably about the unexpected rise of natural gas, but otherwise was the standard anti-Obama boilerplate.

Obama is an easy target for coal moguls, although he's a strange one since he really hasn't made much of a dent in any part of reinvigorated environmental protection, notably curbing greenhouse gas emissions. With an election on the horizon, Obama couldn't afford to take on the mining and utility industries and do much about coal's environmental and safety issues. Critiquing Obama, conservatives such as Peggy Noonan, a *Wall Street Journal* columnist, were smirking that dealing with global warming was just too low a priority to bother with. In her September 24, 2011 article "Amateur Hour of the White House" she wrote, "The decision to focus on health care was the president's own," in a scathing assessment of Obama's presidency. "It could have been even worse. Some staffers advised him—this was just after the American economy lost almost 600,000 jobs in one month—that he should focus on global warming." Sneering about Obama and pressing environmental matters would do much to bring progress on such matters to a crashing halt.

There were other reasons why conference attendees in Palm Beach Gardens were upbeat. The export market, especially for metallurgical coal, was going gangbusters. At the conference, Bob Reilly, the senior vice president for business development at Peabody Energy, spoke first, confirming the optimism. The industry is "in the early stages of a coal supercycle," he said. "The best days are ahead. We have a certain standard of living in the West and Asian nations and the rest of the world want it and they are building out." He declared that "it's going to be a long-term sustainable cycle for us," because about 90 percent of the U.S. coal industry's export growth comes from India and China.

As speakers at the confab confirmed, the coal industry was actually blindsided by a sudden burst of interest in metallurgical coal from Asian nations, which has since moderated somewhat. By 2011, prices had shot up to $330 per metric ton (delivered), a substantial chunk of change. Alpha Natural Resources, for example, got a spectacular $400 a ton for metallurgical coal on the spot market in 2008. "It was great, but it was only for a couple of boatloads," says Crutchfield. Coal-hauling railroads found themselves hard-pressed to handle the new and sudden demand. Norfolk Southern, for instance, needed 2,500 hopper and gondola cars and forty-two new locomotives almost overnight in 2010. Demand has since cooled slightly, but the long-term outlook remains robust. In the United States, the most growth will come from Wyoming's Powder River Basin, teeming with sixty-foot swaths of easy-to-mine, low-heat, but beneficially low-sulfur coal just beneath the butte-dotted grasslands. Older Illinois coalfields are making a comeback. In Central Appalachia, where the production formerly owned by Massey Energy faces the lingering problems of thinner coal seams and more costly and destructive mining practices, the coal is the best in the United States. It is high-heat, and low-sulfur, and has strong qualities to make coke to smelt steel. "We think this is a long-term, sustainable cycle," Reilly said. "These are the new good old days."

Whether that statement holds up depends, of course, on how long countries such as China, India, and Brazil can maintain their economic hot streak. Betting money says they can, at least in the short and middle term. Luke Popovich, a media spokesman at the National Mining Association just a few blocks from the U.S. Capitol in Washington, says that the coal boom will last "as long as Asia enjoys above-trend growth even of 5, 6, or 7 percent a year." In 2010, China and India grew by 10 percent respectively. Even with a slowdown of a few percentage points, we will still be witnessing "the Third Industrial Revolution," says Popovich, who notes that in 2011, U.S. exports of coal will reach 100 million tons, the highest level in forty years.

To see how Central Appalachian coal reaches the global market, I took a trip to Norfolk, Virginia. One cloudy, late-summer morning, the red-hulled *Navios Alegria,* a 738-foot bulk carrier of Panamanian registry owned by Navios MLP, a Greek firm listed in the Marshall Islands, nestled alongside Lamberts Point Pier 6, the northern hemisphere's largest coal pier. Just beyond her on Norfolk's Elizabeth River was docked the *Weser,* a somewhat smaller German collier. The black-painted pier facility is owned by Norfolk Southern railroad, one of the two largest coal haulers in the eastern United States. All day, every day, diesel locomotives push and shove railcars filled with coal to various points of the four-hundred-acre facility, which can handle up to 6,500 railcars. About 95 percent of the coal exported through Lamberts Point is bound for steel mills in Europe, South America, and, more recently, China, according to Robin Chapman, a senior communications official with the railroad.

We drive about the yard in Chapman's small car, viewing a huge rail marshaling yard and then a smaller collection of tracks called a Barney yard. "We have no idea where that name came from," says Chapman, but it's not from the children's-show purple dinosaur. The coal loading method is distinctly low-tech and has been around for decades. There, silver and red hoppers or gondolas are shoved by locomotives to the top of a small hill called a hump. Some of the cars have the lettering TOP GON on them. It comes from a Norfolk Southern marketing effort dating from back in the 1980s to steal some buzz from the hit movie *Top Gun,* featuring Tom Cruise zooming about in a navy F-14 Tomcat jet fighter. One at a time, the "Top Gons" and other, lesser cars are released down the "hump" hill, where gravity takes them to a rotary dumper. This huge device grabs the car with gigantic metal arms. The entire car is then rotated upside down and all the coal is dumped out. The car shuttles up another incline and then zooms back to the yard as if in a theme park ride. The coal is sorted underground according to proprietary codes chosen by the coal firm and shipper. From there it rests at one of two tall towers on Pier 6 with large arms that dump the coal, according to exact size and type, into the holds of the ships.

It takes about thirty hours to load a ship. In the past few years, the demand for metallurgical coal swept up in a surprising wave. One reason was China's demand for steelmaking coal had roiled world markets when Australia, a leading exporter to China, had serious weather problems that scrambled its rail and port infrastructure. That sent Chinese business all over the world to places like Norfolk. It wasn't the biggest burst of export interest. After the Iranian oil crisis in 1979 and 1980, oil prices rose so quickly that utilities in Europe and elsewhere desperately needed steam coal to serve their customers. At that time, as many as one hundred coal ships were swinging at anchor at Hampton Roads, waiting for dock space. The more recent boom hasn't been so dramatic but will likely be more significant and longer lasting. Mark H. Bower, group vice president at Norfolk Southern, noted at the Florida conference that the met coal boom caught Norfolk Southern short. It suddenly needed 2,500 new or leased coal cars to handle the load, and forty-two new locomotives. CSX Transportation, another Eastern coal hauler, also found itself short of rolling stock to rail coal from Central Appalachia to its export facility at Newport News, just a few miles away. Several months later, Bower later told me that the coal car and locomotive situation had been 95 percent rectified and demand had calmed down a bit. Asked if the global demand for met coal will continue, Bower responds that it depends on global prices for coal, which are impossible to predict—as are weather patterns in Australia. Even so, he says, it is looking good.

So good, in fact, that the West Coast is getting in on the act. The rush is on to build new coal-exporting facilities in places like British Columbia and Washington State. They would tap steam coal from the Powder River Basin coalfields of Wyoming and Montana that so far had been used only in U.S. and Canadian utilities. Arch Coal, the nation's second-largest producer, is putting in $25 million for a 38 percent stake in an export terminal on the Columbia River in Washington State. Another U.S. producer, Cloud Peak Energy, is likewise searching for a way to haul coal from the Powder River Basin in Wyoming and Montana. At the Florida conference, Colin Marshall, president and chief executive of Cloud Peak Energy, said his firm is looking into building new coal-loading terminals at the British Columbia ports of Westshore and Ridley. The coal would be railed a lengthy two thousand miles from the firm's Spring Creek Mine in Montana before being dumped into the holds of ships whose destinations are likely to be in Asia. The only caveat: the price of steam coal has to stay relatively high at about eighty dollars per metric ton for the plan to work; otherwise, other suppliers and means of energy production become more cost-effective.

Nuclear power was expected to become the power of choice, given its limited contributions to global warming and its generation capacity to kick out great gobs of power. That is, right up until the Fukushima disaster shattered the hopes for nuclear power, certainly in the short term and possibly forever. Fukushima, a pleasant city of about 300,000, is the capital of the prefecture, or region, where the nuclear power plant is located. The town is about thirty-six miles from the plant, which is surrounded by a closed zone too irradiated for normal travel. About 100,000 people have been evacuated from the closed zone, the central part of which had high levels of gamma rays that can penetrate steel, wood, and human skin and cause cancer or death.

On the day I visit the region, however, radiation seems to be far from the minds of the local people. In a square by a multi-level shopping mall and train station is a holiday festival. A miniature steam locomotive pulls youngsters on a little track as doting grandparents smile from the sidelines. Two young policewomen in sky-blue uniforms lift up children to have their pictures taken on their Honda motorcycles. A brass band, followed by a traditional Japanese music group, blares out show tunes. If the Daiichi plant is of concern, the only clue is in a newspaper article that a man shows me. It has a detailed map showing how many houses have been evacuated. I count as many as seven thousand within the contaminated zone that are too irradiated to occupy. Shortly afterwards, on October 14, 2011 *The New York Times* reports that hot spots of radioactivity have been found in Tokyo, 160 miles to the south.

The Fukushima disaster has been ranked on the level of Chernobyl in terms of radiation released and the amount of equipment melted down. Although nuclear power has been looking like a good alternative for energy production, reactors—most more than thirty years old—are showing their age; and as if to underline the susceptibility of nuclear plants to earthquakes, five months after Fukushima, on August 23, 2011, another earthquake rocked the hopes for nuclear power's revival. This time, the quake was in an unlikely spot: the center of Virginia. It rocked much of the U.S. East Coast, damaging the Washington Monument and raising even more questions about nuclear safety. The quake scrambled power stations far more significantly than expected, dampening hopes of building more of them to replace coal-fired capacity. In one example, Dominion Virginia Power was planning on building a third nuclear unit at its 1,800-megawatt North Anna Power Station, which stands by rolling hills and a large man-made lake in central Virginia. Dominion had been examining new reactor designs. But the August earthquake measuring 5.8 on the Richter scale approached the plant's earthquake stress load of a 6.1 Richter level. For the first time ever in U.S. history, the plant's two reactors were "tripped," or shut down, by the quake. The U.S. Nuclear Regulatory Commission was so concerned after the 2011 quake that it launched an intensive probe of North Anna and other nuclear plants and found that at least 25 percent of them need major design improvements to withstand earthquakes.

Even if the NRC review leads to improvements at the plants, new ones still face more trouble. New plants can cost from $8 billion to $10 billion or more. Costs that big mean that utilities will need federal loan guarantees to get credit from the private sector. Yet the financial community, still traumatized after its own crisis in 2008, is wary of lending that kind of money without guarantees. Congress, likewise, is under such strong pressure to cut spending that it likely won't do much to underwrite new plants.

Other countries have taken steps to curtail nuclear power generation. Chancellor Angela Merkel of Germany announced that her country would shut down its nuclear plants by 2022. Switzerland made a similar announcement. Siemens, the huge German conglomerate, also announced it was getting out of the nuclear business. Areva SA, the French firm that is the world's largest supplier of nuclear services and fuel, was considering writing down the value of its uranium. This remarkable and unexpected turn of events has uprooted thinking that nuclear power has proved itself as the safe, reliable way to displace fossil fuels for energy needs.

The mayhem over nuclear power gives coal a boost, but so does another turn of events. Coal has been under strong attack for the monumental amount of greenhouse gases it produces, which contributes to climate change. One special problem with coal is that when it is burned, it gives off more carbon than other forms of fossil fuel such as oil, natural gas, and other petroleum derivatives. Of all fuels coming from decayed plants or long-dead dinosaurs, it is by far the most polluting and the biggest contributor to greenhouse gases because of its extra portion of carbon, which ends up as carbon dioxide.

According to author James Fallows, who wrote the influential article "Dirty Coal, Clean Future" in December 2010 *The Atlantic* magazine, all human activity puts about 37 billion tons of carbon dioxide into the air each year. Twenty-five years ago, the carbon dioxide rate per year was 25 billion tons, and within twenty years, it could be 50 billion tons a year showing that the trajectory is to double carbon dioxide emissions within a half a century. The biggest polluters by far are Americans, who account for twenty-five tons per person per year. Europeans are considerably less polluting, averaging only eleven tons per person, and the Japanese are slightly less than that. China is lower still—eight tons per person—but that will climb as the Middle Kingdom embraces the materialism of its emerging middle class.

For years, engineers have been studying how to reduce the carbon signature so that coal pollutes less. Any number of possibilities exists to eliminate carbon dioxide and other pollutants before or after the coal is burned. Coal can be pulverized to such a degree that the resulting powder burns very cleanly. It can be burned over a bed of limestone that can trap much of the resulting sulfur and some of the carbon. Postburn technologies involve capturing the carbon before transferring it to permanent storage areas far below the earth's surface. Chilled ammonia can be used to trap carbon before it goes up smokestacks. These are just a few methods.

Yet the movement to do much about such gases has been stalling. Experts agree that the United States is behind Europe and Asia in the coal-related carbon technology. One reason is that Congress has moved only in fits and starts to realize the importance of it. Another is that the coal industry has used its deep pockets to lobby legislators and regulators and propagandize the public to limit any new regulations on their products or do much to limit their top line and net profit margins.

The electric utility industry seems split, even among big coal-fired ones. American Electric Power, based in Columbus, Ohio, is the utility most dependent upon coal. It is located near the heart of the Appalachian and Illinois Basin coalfields, giving it a major cost advantage in electricity production. It has blunted efforts to restrict coal use, although it has dabbled in carbon-capture technology. Duke Energy, based in Charlotte, North Carolina, is likewise coal dependent and is not too far from important coalfields, but it has taken a more proactive view on greenhouse gas emissions. Utilities like Duke had backed a cap-and-trade market concept to give utilities so many carbon "credits" that could be exchanged or purchased. The scheme tended to benefit them at the expense of coal producers because it gave utilities more flexibility to trade. Regardless, the cap-and-trade concept was all but abandoned given more pressing immediate issues such as unemployment, a struggling economy, and the newfound fire among conservatives on deficit spending. Thus, the stage is set for a conundrum that ruins prospects for significant progress on carbon-capture technology and fewer green-houses gases. The key is passing federal legislation requiring new carbon rules. The industry won't do that by itself. Indeed, it probably can't do it alone, since it is under so much pressure from Wall Street and its shareholders to produce dividends and profits. Old-style, baseload utilities that have large coal-fired or nuclear plants that provide the bulk of electrical energy are beholden to state regulators who set electricity rates. Typically, regulators find it difficult to approve rate hikes for experimental projects to snare carbon at coal plants since they are required by law to grant hikes only for production methods that have the cheapest costs.

One telling example is the Mountaineer Power Plant in Mason County, West Virginia. Built on the banks of the Ohio River in 1980 by AEP, the 1,420-megawatt plant, boasting one of the tallest smokestacks in the world, was picked for a pilot program to determine if new technologies can be retrofitted on existing plants. The idea was to use chilled ammonia to capture carbon so that it isn't emitted to the atmosphere. The carbon would then be stored and injected deep into the earth under a technology developed by Alstom, a French engineering firm that specializes in developing various carbon-capture methods. The plan was to remove up to 300,000 metric tons of carbon dioxide from the 8.5 million tons emitted each year at Mountaineer.

What killed the project is who pays for it. The price tag was initially $100 million and would go to $668 million with a pledge from the U.S. Department of Energy to fund half the cost. That high price tag and political shifts proved its undoing. In July, AEP announced it was suspending the project because regulators in Virginia and West Virginia refused to include the pilot project in the utilities rate base and have consumers pay for it. Utility officials likewise were afraid to depend on a DOE pledge for half the cost after Republicans won the House of Representatives in 2010 and created a new climate of federal cost cuts. The lesson is that carbon-capture technology won't happen unless there is a federal law requiring it. Ned Leonard, vice president for the American Coalition for Clean Coal Electricity, an industry lobby, told me that, in theory, chilled-ammonia technologies can capture 90 percent of carbon dioxide, but the maze of funding problems is holding it back. "AEP has to recover its costs," Leonard says. "But when the utility went to regulators for permission to pass costs on to consumers, they ran into the problem of the 'least-cost power' policy. The regulators said no because law compels them to."

Mountaineer is not the only example of a carbon-capture project stumbling because of anemic, post-recession finances. In 2003, the Bush administration pushed the idea of a power station that would turn coal into a gas and then burn its hydrogen for power. The Obama administration tweaked the idea so that coal would be burned in pure oxygen instead of in ordinary air, so that the resulting carbon dioxide, a greenhouse gas, would itself be in a purer form. Obama's stimulus project gave the plant, located in Meredosia, Illinois, $1 billion to go forward. In November 2011, Ameren, the utility that was to host the project, announced it might not be able to proceed because of its precarious financial situation.

Stumbling demonstration plants show just how far President Barack Obama's early ambitions for cutting greenhouse gases have fallen. In his first year in office, he pushed a goal of an 80 percent reduction in greenhouse gases by 2050. He did try to push for alternative energy forms, such as solar panels. In that effort, the federal government gave loan guarantees worth $535 million to a California-based solar panel maker called Solyndra. Praising the start-up company as a step toward "a brighter and more prosperous future," Obama made Solyndra a poster child for his efforts to push green energy and kick-start job creation. The firm, however, went bankrupt in 2011, taking with it the taxpayers' loan guarantee and 1,100 jobs. Obama opponents had a field day mocking the project, which represented only 2 percent of about $40 billion in such loan guarantees. Other ones, such as manufacturing automobile batteries, have worked. Unfortunately, federal backing for carbon-capture facilities would have to involve much bigger loans and may not create many jobs other than temporary construction work retrofitting older plants. As the Mountaineer project shows, they won't be built anytime soon.

As efforts to deal with global warming stall, the related political arena has taken some bizarre new turns. Hard-right politicians and global warming deniers are shifting the argument, claiming that maybe humans have no relation to and deserve no blame for climate change. Plus, they are taking legal action and implicating scientists who say humans do have something to do with it. One case involves Michael Mann, a highly regarded climatologist at Penn State University who has worked previously as a researcher and professor at the University of Virginia. Mann has been the object of a smear campaign led by hard-right conservatives who, to the stunned dismay of scientists and politicians around the world, want to challenge the widely accepted view that humans have anything to do with global warming. They went on a publicity campaign when there was a flap at the University of East Anglia in eastern England, which is a repository of global warming documents for the United Nations. London newspapers screeched over a supposed scandal in which an East Anglia professor supposedly suppressed data raising questions about humans and their impact on global warming. Mann spent months defending himself against the assault. After the University of Virginia fought back one set of legal demands for Mann's documents, other global warming deniers would step in and pursue them through the Freedom of Information Act. On March 2, 2012, the Virginia Supreme Court turned aside the demands for records made by Kenneth Cuccinelli II, Virginia's attorney general, but others still pursue them through the Freedom of Information Act.

In November 2011, the U.S. Department of Energy reported that the global output of carbon dioxide had jumped by the highest levels ever recorded, and little is being done to stem the damage from greenhouse gases. Coupled with the carbon dioxide has come what many scientists say is unmistakable changes in weather patterns. The United States has seen a number of record-breaking extremes, including massive outbreaks of tornados and dogged droughts in places such as Texas. The more enlightened leaders in Big Coal may pay lip service to global change, but maintaining the status quo is more to their advantage.

Although the economic recession has temporarily flattened demand for more electricity in the United States, there is no question that more generation sources will be needed. In 2007, just before the recession, the North American Electric Reliability Corporation was projecting that electricity usage in the United States would grow more than twice as fast as resources to supply it by 2017. Older generating units will have to be refitted or shut down. Most nuclear power stations were built before 1980. Adding new transmission lines to carry electricity over long distances lags behind projected demand. If that weren't enough, the workforce running power plants and switches and fixing damaged lines is aging. In 2009, about 40 percent of that labor force became old enough to retire, NAERC says.

The new demand for growth isn't necessarily coming from a boom in new babies but the increasing use of electronic devices brought about by today's information and digital ages.

Cable boxes that give access to hundreds of television and movie channels are huge energy hogs. Consumers tend to leave them on all day and night, further draining power. Desktops, laptops, wide-screen televisions, iPods, cell phone chargers, home Wi-Fi receptors, and various other devices also demand power. And if a revolution in automobiles comes with the widespread use of battery-powered vehicles, those fancy new cars will have to be plugged in somewhere to recharge.

Besides being a huge burden on future electricity generation, technology also could be part of the answer, says David Hawkins of the Natural Resources Defense Council. New ideas are evolving to integrate electrical grids so new and alternative sources of power can be tapped and used when needed. These could be solar farms or even individual panels on a homeowner's roof. The idea is the antithesis of the concept of baseload generation, and Hawkins says it could be used in much the way the Internet is—millions of small generations. Another promising trend involves producing electronic devices that use far less than their electron-gobbling predecessors. "Big flatscreen TVs use a lot, but LED screens use a lot less. The conventional problem is that you have a lot of boxes sucking power without doing anything useful. There are huge improvements coming, and it will come as older consumer electronics get replaced by more advanced versions," he says.

Hawkins acknowledges that coal is still here to stay, at least for a while, even though he says it contributes to the deaths of ten thousand Americans every year through lung disease caused by coal-related air pollution and other issues. Coal's proponents, such as Ned Leonard of the American Coalition for Clean Coal Electricity, counter that coal is very valuable since there is so much of it in the United States and China. "Our country has enough coal to last two hundred fifty years," he says.

One new competitor is shale gas or natural gas drilling through new techniques called hydraulic fracking. First put in wide use in western New York State and western Pennsylvania, fracking involves nudging out hard-to-get layers of natural gas by using chemicals and high-pressure water injection. Results have been so successful that Leonard says estimates of the U.S. natural gas reserves are soaring. "They used to say we had only a fifteen-year reserve, but now they say a hundred and fifty years," he says. Yet fracking is controversial since the chemicals involved can be highly toxic and the depths at which the drills operate are so deep that much isn't known about the lasting impact. While they supposedly go below groundwater tables that supply wells with drinking water, critics say they are threatening that vital resource.

Coal-rich states such as West Virginia are looking at fracking, and traditional natural gas drilling has been going on for decades, but the areas most likely for fracking are not in the same places where the best coal reserves are, namely, those owned by Massey and now Alpha. But the sudden and surprising success of natural gas has startled coal executives. "Even in the absence of climate change legislation, thermal coal is facing into a steady headwind, and that headwind is cheap natural gas," says Alpha's chief executive, Kevin Crutchfield. "It's been a very disruptive force on the marketplace. Coal has come out of that battle with a bloody nose."

Gas's success isn't just its cheap price, which can change as quickly as coal's can. Utility executives are increasingly choosing natural gas as the default fuel since there is no clear-burning coal. In addition, retrofitting existing coal-burning plants is more expensive than building new ones with more modern technology, so modifications to make plants less polluting are not being made. So, rather than invest in expensive technologies that may or may not be needed to keep using coal, they are opting for cleaner natural gas.

Hydro-fracking is also flooding the energy market with cheap natural gas, but it still leaves many questions about the toxic chemicals used in the drilling process and whether it affects water tables. Fracking has been blamed for some unusual earthquakes in Ohio, raising the specter that its full impact won't be known for some time. So it may be too soon to write coal off as a major source of fuel for electricity. The Florida conference closed on a high note for coal production in general, especially for coalfields in the rolling badlands of Wyoming and in the Illinois Basin east of St. Louis. And what about the coalfields of Central Appalachia? Despite their uncommonly rich product, they remain a big question mark since the easiest- and cheapest-to-mine seams are being used up, and price futures have to be at high levels to make it worthwhile to rip apart thousands of acres of mountaintops or risk the lives of miners deep below the surface.

Another benefit for coal is that it still is more reliable than so-called alternative forms of energy. Wind, solar, geothermal, and hydroelectric power all have special appeal, because they don't pollute. The technology exists to make them all viable contributors, says Hawkins of the National Resources Defense Council. Yet there are significant limits. At best, wind could perhaps make up 20 percent of electrical production by 2030, according to the Department of Energy. Google, which uses tremendous amounts of electricity to run its servers and air conditioners at a dozen or so "farms" to keep its search engine going, is so worried about future power availability that it is taking matters into its own hands. Google has opened a server farm in chilly Hamina, Finland, where electricity costs are cheap. Facebook is likewise looking at a facility in equally cold Luleå, Sweden, located on the same latitude as Fairbanks, Alaska. Going a step further, Google is also planning its own offshore wind farms. One such project is the $5 billion proposed Atlantic Wind Connection, which would set up an enormous line of wind turbines in the Atlantic some miles off the Mid-Atlantic coast roughly from Norfolk, Virginia, to northern New Jersey. Google Energy, a subsidiary of the Mountain View, California, mother company, will install a seabed infrastructure of cables and connectors on which wind turbines would be fixed. The project would eventually produce 6,000 megawatts of power.

The amount may sound impressive, but in the larger scheme of things, it is a drop in the generation bucket. The amount is about the size of perhaps five large coal-fired plants. After American Electric Power, the country's largest coal-fired utility, takes off about a dozen coal plants that are too old to meet new EPA carbon regulations, it will remove about 6,000 megawatts from its production capacity. In that case, Google's contribution to overall electrical capacity will merely be a wash, pointing out just how limited alternative sources still are. Solar, geothermal, and other energies have similar limitations, although the NRDC's Hawkins says that new technologies should improve the impact and production capacities of alternative energy forms.

Massey Energy also didn't get much attention at the conference. At cocktail receptions, bringing up Massey's problems was like talking about an alcoholic uncle. Attendees seemed to want to forget all about Don Blankenship and his firm. One speaker framed the new thinking best. Nick Carter, who is the president and chief operating officer of Natural Resource Partners in Huntington, West Virginia, and is regarded as the Grand Old Man of Coal, said, "Ten years ago, the top ten [coal companies] produced forty percent of the coal, and now they produce sixty percent. Since more coal firms are publicly traded, we are a much more responsible industry and are much more responsive to the public."

Carter's last statement carries one big irony: Massey Energy was being publicly traded during the years it was building its notoriety for cutting costs, skimping on safety, and intimidating anyone in its path. The list included institutional investors, "green" stock activists, and individual experts in the coal industry. Massey was famous for controlling its shareholders' meetings, held occasionally at the four-star Jefferson Hotel in downtown Richmond, where local police on horseback, muscular hired security guards in dark suits, and smooth public relations consultants kept critics at bay.

The last laugh may be Blankenship's, however. Even as scientists, activists, and politicians note that the burning of coal generates enormous amounts of greenhouse gases, and they decry the wholesale destruction of the Appalachian mountain landscape to make mining thin seams of rich coal more cost-efficient, the fact is that popular alternative energy forms, such as wind or biomass, pollute less but can't achieve the economies of scale that a big coal-fired plant can, and nothing substantial has been done in this country to address dirty coal's problems.

6

STRIP-MINING ON STEROIDS

Vernon Haltom could be a hippie actor in 1960s-era guerrilla theater. His stringy, salt-and-pepper hair is shoulder length, and his round wire-rimmed glasses are refugees from the days of John Lennon. The rest of his attire, however, is ultrarespectable: black suit with muted tie and a starched white dress shirt with navy stripes. He sits in the waiting room of the magistrate court at 115 West Prince Street in Beckley, West Virginia, amid a few pieces of beaten-up furniture, a Dr Pepper machine, and a Lance snack dispenser. It is a place for people who have had scrapes with the law. A young blond woman is crying into her cell phone as she pleads for an update about her child, whom her boyfriend took away in his pickup truck. Her mother sits patiently next to her.

"I've been here before," says Haltom, his moon face beaming in an impish smile. He has, many times. On this occasion, he has been summoned to appear before a judge to explain his arrest two months before by a West Virginia state trooper. The executive director of the eco-activist organization Coal River Mountain Watch, Haltom was part of a unit that marched up to a guard shack of Massey Energy's Marfork Mine to present a written plan for transforming the firm's massive mountaintop-removal surface mine into a line of power-generating and far more eco-friendly windmills. He and his band of eco-pranksters chose the Marfork Mine near the town of Pettus on Route 3 in the Coal River Valley because it is right across a small bridge and a railroad track from some state police barracks. That way, Massey's security guards would have an easy place to get cops to round them up. It was a clever diversion. While the security guards were distracted by the arrests of Haltom and his entourage for trespassing, another group of five activists, the real strike force, sneaked atop the mountain and chained themselves to huge earth-moving machines. The military precision of the operation should not be surprising. In an earlier life, Haltom was an officer in the air force charged with guarding nuclear weapons.

After waiting at the magistrate court for more than two hours, Haltom is released. The state trooper failed to show up. It had been another of many courtroom episodes for Haltom and his organization, which is based on the first floor of a building in Whitesville, a small town in the center of the Coal River Valley about forty miles from Beckley. It is just a few miles from two gigantic mountaintop-removal operations owned by Massey Energy.

Later, the Coal River Mountain Watch headquarters is buzzing with half a dozen young people peering into Apple laptops or trying to get the coffeemaker to work. Filling part of one wall is a child's big painting in bright green and brown depicting windmills on one side of a big mountain with a strip mine on the other. A number of black-and-white NO COAL PLANT signs used to protest electric utilities' building of new coal-fired generating stations also dot the wall, along with bumper stickers stating I ♥ MOUNTAINS. Their pro-coal opponents in the propaganda war have their own, nearly identical bumper sticker stating I ♥ MOUNTAINS & PLATEAUS. The "plateaus" allude to supposedly beneficial reclaimed surface-mine land flat enough to be used for buildings.

Hopes had been high, Haltom says, that Barack Obama might tighten policies and bring some sanity to mountaintop removal that has destroyed thousands of acres of wildlife habitat, trees, and bushes; ruined hundreds of streams; and polluted groundwater. "He seemed to have considered it for a while," Haltom says. "But he has to face the West Virginia's and Kentucky's powerful political forces that back coal, including [West Virginia's U.S. senators] Manchin and Rockefeller and the Tea Parties."

Oklahoma-born Haltom arrived in the Coal River Valley about ten years ago when his wife, Sarah, and a native of the area, wanted to return home. Besides pursuing activism, he teaches at local colleges. The group was led by Julia "Judy" Bonds, a southern West Virginia woman who sprang into action in 1996 when her six-year-old grandson began scooping up handfuls of dead fish in a creek near land that had been in her family for generations and where she had fished for decades. "What's wrong with these fish?" the child asked. The answer was the Marfork Mine, a surface operation that had begun in the 1970s but greatly accelerated when Massey bought it. A company that she believed "has the most unconscionable record in Appalachia."

The fish kill was a moment of clarity for Bonds, who had been working as a waitress at a Pizza Hut and at convenience stores. She became the godmother of the movement against mountaintop removal. Given her hillbilly pedigree as the daughter of a coal miner, it was hard for the public relations men of Massey and other big coal companies to depict her as an elitist dilettante. Contacting thousands of people, she built up the Coal River Mountain Watch, which paid her twelve thousand dollars a year, and formed global networks with other groups, such as the Ohio Valley Environmental Coalition, the Sierra Club, and the Rainforest Action Network, that pressure the institutional investment community not to lend money to errant coal firms. In 2003, she won the prestigious Goldman Environmental Prize, spending the $125,000 award on her grandson's braces, helping her daughter pay off her car, paying off her mortgage, and giving $50,000 to her group. Protests against mountaintop removal have drawn a galaxy of concerned celebrities, such as Daryl Hannah, the half-fish, half-woman star of the Ron Howard movie *Splash;* Nashville country singer Kathy Mattea; and Robert F. Kennedy Jr., an environmental lawyer and son of the former senator and assassinated presidential candidate.

When Bonds died of lung cancer in early 2011 at the age of fifty-eight, Haltom took over as executive director. From its Whitesville base, the group keeps up its work educating the public about the dangers of mountaintop removal and is watching to see what happens as Alpha Natural Resources assumes Massey's mountaintop-removal operations. Locally, the horror of the Upper Big Branch disaster has surpassed surface mining as a major cause célèbre. Nationally, however, Coal River Mountain Watch still remains a focal point for information about mountaintop removal for environmental groups.

To understand why mountaintop removal spurs so much outrage, it has to be seen from a bird's-eye view. Driving up and down the winding country roads won't work, because the steep hills block views, and most highways run along low-slung creeks or rivers. The only way to see mountaintop removal in its entirety is from an airplane. One way, although fleeting, is on a commercial jetliner at thirty-five thousand feet on a flight about a half hour out of Dulles International Airport near Washington, D.C. In summer, dense dark-green forests suddenly give way to clearly visible tan and dark-brown splotches that seem unusually large and continue for perhaps ten minutes. It is as if a tropical jungle had suddenly changed into the barren, alkaline wastes of the Nevada desert.

A better way to get a feel for their huge size is by flying at low altitude in a smaller propeller aircraft or a helicopter. Few people manage or have cause to do this, but if more did, the outcry against mountaintop removal might be much greater. Like so many of the injustices of Central Appalachia, the ravaging scars from the mining practice are hidden. People living just a few hours away by car in their suburbs or cities don't know what is going on, but this has been the story of the mountains for decades.

Having seen smaller forms of strip-mining when I was a boy and having viewed mountaintop removal from the ground, I naturally wanted to see it from the sky as well. The pilot at the Beckley airport is a young man named Kevin who wears a striped polo shirt, shorts, and sandals. He fits Scott, my photographer, in the front seat and me in the rear of a thirty-one-year-old Cessna 172. "It may be bumpy," warns Kevin as we roar off into the late-spring afternoon.

Soaring west, we float over beautiful gorges and crenulated mountains that look like giant brain coral. At about four thousand feet, we are around Massey's Edwight and Marfork mines to the west and east of Coal River Valley. Kevin startles us by jamming open Scott's window in a quick motion. He keeps it open by placing a small metal strut between the window and the fuselage so Scott can take his pictures. Wind howls in the cabin. "Don't lose ma' stick!" he shouts to Scott.

The mountaintop mine site is unmistakable. Its proportions are oversized and spectacular, running perhaps three by four miles. The cutaway area stretches the size of perhaps twenty large-scale shopping malls, including parking lots. Yet the surface mine is multilevel and multidimensional. Big coal trucks about three stories tall run up and down dirt roads that dogleg among stripped-away areas. A gigantic blue and white dragline, or electric-powered earthmover, shovels away scores of tons of coal in a single scoop. The scene presents objects so utterly massive that my eye automatically scans the ground for some familiar object so I can get a sense of scale. There are some trailers that must be offices for managers and some high-tension power lines to keep the shovel running. They're tiny in comparison. In the distance are reclaimed areas, but the contrast between their light-green grass and the darker shades of still-untouched hardwood trees in natural hills can't be missed. The mine sites spill over so many hills and valleys that it takes us a good five or six minutes of flying just to make a turn and get past it.

Although I have seen mountaintop sites before, they have always been from the ground. When I consider the comparatively tiny abandoned strip-mine sites I used to play on in the 1960s, the vastness of the aerial view I have just witnessed is personally overwhelming. Scott and I banter with the pilot as best we can as he secures the Cessna, but we go quiet as we drive away.

Although the prize is coal, strip-mining has its roots with the internal combustion engine. Large building projects that had been completed previous to such engines, such as the Panama Canal, had been severely hampered because mechanical shovels had been powered by inefficient and ungainly steam engines. The evolution of the gasoline and, more important, the diesel engine changed that, but even when earthmoving equipment such as bulldozers, earth scrapers, draglines, dump trucks, and tanker vehicles were driven by combustion engines, they didn't see much use except on a few big projects such as the Hoover Dam or in the creation of a series of lakes and dams for the Tennessee Valley Authority during the Depression. It was only after World War II that such machines found their way to the Appalachian Mountains.

Hundreds of thousands of armored tanks ran on gasoline or diesel fuel as they rolled through the plains of Ukraine or the sands of the Sahara in Northern Africa. In the Pacific, petroleum-powered earthmoving machines were high-priority items for building forward staging bases and airports in the island-hopping campaign against Japan. Thousands of U.S. Marines died seizing islands such as the Marianas so coral could be bulldozed for giant runways capable of handling B-29 bombers that would later incinerate Japanese cities in raids. The two bombers that dropped atomic bombs on Hiroshima and Nagasaki embarked on their historic missions from Tinian, where bulldozers had hacked out lengthy runways.

When the war ended in 1945, large numbers of surplus bulldozers, trucks, and draglines went on sale to civilians. While strip-mining had become common in the 1930s, it really took off after surplus war gear became available for independent coal operators in the Appalachians who had worked smaller deep mines on a contract basis. Larger coal firms snatched up the gear, used and new, as well, but the independent operators really saw a way to make money, especially in parts of the mountains where coal seams had run their course with deep-mining but where there were leftover parts of the seams near the surface, because those tab ends could easily and cheaply be exploited with bulldozers, explosives, draglines, and trucks.

Soon after, strip-mining began in earnest, often by small crews known as "shoot-and-scoot" operators. They blasted away rock to get at the coal and then left quickly without any reclamation. It seemed a great new way to make money from coal. Surplus equipment was cheap and plentiful. Much more could be done with fewer workers. The operations were a fraction of what deep mines cost, and although miners could be fatally injured in blasts to loosen rock, or electrocuted or killed if trucks rolled over, surface mining was far safer than working underground. With a couple of bulldozers, an air compressor to drill holes in the side of the hill to place explosives, and a dragline to scoop up the rock overburden and attack the coal seam, only four men were needed to operate these bare-bones mining operations.

The bonanza was on. In short order, stripped-out brown gashes, called "highwalls" and sometimes ninety feet or more in height and running for tens of miles, marred the sides of Appalachian hills from Pennsylvania south to Alabama. At first, the coal seams tapped were those near the sides of hills that had been left over from underground mining because they were too expensive to dig out. The process was simple: Surface miners bulldozed out an earthen shelf and access roads. All vegetation, sometimes including virgin forest, was cut down. Augers bored into the coal seam, and explosives were inserted. Debris from the blasts was hauled away, and the bulldozers, draglines, and more augers got to work loosening the coal, which the trucks hauled away to preparation plants or rail or barge load-outs.

The exceptional ecological devastation of the practice soon became apparent. In the early days, many of the mines were simply left without reclamation. The "overburden," or resulting waste from the chopped-away tops or sides of mountains, was pushed over the side of the shelf. Wildlife habitat was decimated. Without trees and bushes to hold it back, rainwater rushed down the hills in greater volumes and velocities, leading to devastating flash floods. Depending on the characteristics of the coal seam being mined, residual sulfur,

Mountaintop removal practiced by such coal companies as Massey Energy involves blasting and bulldozing away thousands of acres of hilltops and putting the waste in streams and valleys, such as at this mine near Edwight, West Virginia.

Above: Coal can be profitable for firm owners such as E. Morgan Massey, former head of Massey Energy's predecessor. He owns and flies these Citation corporate jets.

Below: Donald L. Blankenship, Massey Energy CEO and chairman, appears before the Senate Appropriations Subcommittee on Labor, Health and Human Services, Education, and Related Agencies hearing on mine safety forty-five days after the deadly blast at the Upper Big Branch Mine on April 5, 2010.

Above left: Poverty remains a devastating problem in Central Appalachia as more than one thousand people flock to this free dental and medical clinic in Pikeville, Kentucky, in June 2011.

Above right: Wearing overalls with flourescent stripes, these miners work for Alpha Natural Resources which took over Massey Energy on June 1, 2011.

Below: Much of the coal in these hopper cars at Williamson, West Virginia, will be exported to steel mills in China, India, and Europe.

Above: At his anti-union 2009 Labor Day rally held in Holden, West Virginia, featuring music stars Ted Nugent and Hank Williams, Jr., Don Blankenship said that officials worried about mine safety were as "silly as global warming."

Below: Environmental activist Robert F. Kennedy Jr. speaks at a June 11, 2011 rally commemorating the Battle of Blair Mountain.

Above: West Virginia State Police separate environmentalists from pro-coal advocates at the Blair Mountain rally.

Below: Local sentiment: A coal miner used his house to protest the environmentalists at the Blair Mountain rally.

Above: Disabled veteran miner Gary Quarles holds a helmet commemorating his son, Gary Wayne "Spanky," killed at Upper Big Branch on April 5, 2010.

Below: West Virginia Route 3 is dotted with memorials to the twenty-nine dead at Upper Big Branch.

This memorial was placed just outside the Upper Big Branch mine.

Miner Tommy Davis of Dawes, West Virginia, barely made it out of the Upper Big Branch as a massive explosion killed his son, brother, and nephew. He hangs his son's workshirt on the flagpole as a memorial.

selenium, and other harmful or carcinogenic materials were exposed and mixed with rainwater. The goo ended up in ponds that formed naturally on depressions on the mine shelf and often took on a sickly orange-yellow color. Since the sulfur and other pollutants were toxic, and the shores of the ponds often were lined with the skeletons of dead animals that had used the ponds for drinking water. The pollution eventually ended in watersheds, ruining creeks and killing fish, frogs, and other animal and plant life.

Henry M. Caudill notes in *Night Comes to the Cumberlands* that in the 1950s and 1960s, strip-mining also grew in other countries, such as the United Kingdom, Germany, and (the former) Czechoslovakia, but their approach was more careful and considerate of the environment. Layers of rock and dirt were laid near the coalface in the order in which they had been removed, including as much vegetation as possible. After the coal had been extracted, the layers were returned in the order in which they had been taken out. Although it was impossible to return the topsoil and vegetation layers to their original states and this was therefore still a flawed technique, it was less harmful than what was being done in Appalachia.

In states such as West Virginia and Kentucky, coal interests dominated the political and legal processes. They bankrolled their candidates in state legislatures and had legions of lawyers and lobbyists to fight regulations. Going the European route was not possible. Coal operators didn't want to be bothered with the time or expense. Another harmful aspect was that surface mine operators often got mineral rights to coal that had been signed away as long ago as the late nineteenth century when uninformed farmers signed away their rights for a few dollars, not quite understanding that the impact on the surface where they lived, grew crops, or had their livestock could be enormous. Of course, in the late 1800s, when the sharply dressed city boys from Pittsburgh or Roanoke or Knoxville picked up minerals rights for a song, no one could have known that diesel-powered earthmovers would be dozing off cherished property that had been in families for generations. It was assumed that all mining would be underground and out of sight.

Within a few years, the devastation became more apparent. Caudill wrote of what happened in Eastern Kentucky:

The community of Upper Beefhide Creek in Letcher County was stripped between 1950 and 1954. Highwalls ten miles long and averaging forty feet in height were created. Earth and rock were tumbled down the mountainsides and raced through fields of grass and corn and blanketed lawns and vegetable gardens. Stones as big as bushels were blasted high in the air like giant mortar shells, to rain down on residences, roads, and corn land. One huge rock crashed through the roof of a house, struck a bed dead center and carried the spring and mattress into the earth under the foundation. Another fell on a cemetery, smashing a tombstone and crushing the coffin six feet underground.

Landowners had little recourse in confronting the coal operators, who left quickly after their jobs were finished and were protected by a maze of alphabet soup company names. They typically did not have the financial resources to hire legal help to sort them out, much less mount court challenges, and as they woke up to find that their cherished landscapes or pastures were being ripped apart and not put back, despite state laws requiring so, they discovered they had another big problem from strip-mining. Stuck with dozens, if not hundreds, of acres of unreclaimed land, their property values plummeted, as did tax assessments, meaning that less money went to their counties or states to provide badly needed services such as public school education, roads, and social services.

The ravages of Appalachian strip-mining remained mostly off the national radar screen until the 1960s. Regional newspapers such as the Louisville *Courier-Journal* started reporting their hazards. One seminal moment came in November 1965, when a sixty-one-year-old Kentucky woman, Ollie Combs, known as the Widow Combs, was photographed being manhandled and lugged away from her Knott County farm by two Kentucky state policemen. Strip-mining operators had threatened her property, where she quilted and raised chickens. Protesting, she refused to leave and was forcibly removed. She spent Thanksgiving Day 1965 in the county jail.

"I have never been in trouble," she told me in 1997 for a *Business-Week* article. "I just want to live my life in my hollow and be left alone." She and her two sons were fed a chicken dumpling holiday meal in jail, and she was released. After her photograph was circulated around the world, she became an anti-strip-mining spokeswoman and influenced the Kentucky legislature to pass tougher surface-mining laws in 1966. With her uncompromising and fearless stances against coal operators and the politicians who backed them, Combs, who died in 1991 at the age of eighty-seven, was the exception to the rule.

Strip-mining continued unabated into the 1970s until tragedy again forced the first serious regulations. In Kistler, a small West Virginia town just up the road from a crossroads village called Man on Buffalo Creek, the weather was spitting sleet on February 26, 1972. Glenna Wiley was in her house just yards from the narrow creek meandering up a hollow past fifteen other hamlets and several coal mines in Logan County. Her telephone rang. An excited neighbor warned that an earthen dam up the creek containing coal sludge from surface mines had burst.

Wiley ran to her porch and looked up the creek. A tsunami-like wall of black water, thirty feet high, was racing toward her. She said she saw the face of another neighbor turn red, then white. As the waves crashed past, she saw a number of human bodies float by. Of the 5,000 people who lived up and down that part of Buffalo Creek, 125 would die, 1,121 suffered injuries, and 4,000 would be left homeless after 507 residences were destroyed.

The disaster was touched off when three earthen dams holding back 132 million gallons of filthy water from surface mine sludge breached. The dams had been built by Pittston Coal Company, most of whose properties were later bought by Alpha Natural Resources. The third dam built by Pittston had been constructed on unstable coal slurry sediment instead of bedrock. Still, a federal mine inspector had checked the dam and declared it safe. Then it had rained for four days, and Dam No. 3 broke on February 26, its liquid spilling over into the other two ponds, breaking their dams as well.

The aftermath brought a mixed resolution for Pittston and other coal operators. Pittston managed to slip past substantial legal penalties. Six hundred and twenty-five survivors sued for $64 million, but settled for just $13.5 million or about thirteen thousand dollars for each plaintiff after legal costs. The state sued for $100 million in relief and damages, but Governor Arch A. Moore Jr., a politician generously supported in campaign contributions by the coal industry, settled for only $1 million. On the other hand, the Buffalo Creek tragedy got widespread attention both in the news and after Gerald M. Stern, a lawyer for Arnold & Porter, the plaintiffs' representative, wrote a book about it. Two laws resulted. The Dam Control Act was passed by the West Virginia legislature in 1973, which intended to stiffen technical construction requirements for dams, but adequate funds were not initially approved for enforcement. It took Congress to pass a much further-reaching surface mine law in 1977 that actually had some teeth. It created the Office of Surface Mining under the U.S. Department of the Interior, toughened regulations for digging and operating the mines, and forced operators to reclaim the land around the mine and return the "overburden" rock and soil to the original contour of the land. Funds were also collected to repair older strip mines where the highwalls and shelves had been abandoned for decades. While that helped to correct the endless gashes along hundreds of hillsides, the pushed-back land was graded and replanted with grass and trees, which never seemed to grow well because the topsoil had been destroyed, and worse yet, the respite would be short-lived. As old abuses of surface mining started vanishing in Appalachia, new and far more formidable ones emerged. Mountaintop removal started in the 1980s and took off in the 1990s, making surface mining methods seem puny by comparison. Driving the practice were technology and economics. In the former case, earthmoving equipment had evolved to a point that was unimaginable a few decades before. Much larger earthmoving machinery had been developed at Wyoming and Montana's Powder River Basin. Gone were the couple of dozers and an aftermarket dragline. In their place were electrified power shovels as tall as twenty-seven-story buildings, weighing eight million pounds, and capable of moving thousands of cubic yards of dirt or coal in one enormous scoop. The mines were no longer dozens of acres but thousands of acres in size. Explosives had been developed that went far beyond the simple dynamite charges used in the 1950s. Today's charges can contain any number of modern explosives, typically ammonium nitrate and nitromethane with an explosive trigger made of Tovex. These were the same ingredients in the bomb that was used by domestic terrorist Timothy McVeigh to blow up the Alfred P. Murrah Federal Building in Oklahoma City on April 15, 1995, killing 168 people. The ones used for mountaintop removal, however, are several times more powerful.

Economics drive mountaintop-removal mining because it still is cheaper to buy gigantic earthmoving gear and take a year to chop down an eight-hundred-foot-tall section of a mountain to reach a seam of coal only a few feet thick. It is certainly cheaper than erecting a complex deep mine if the seams are thin, as many of the best ones in Central Appalachia are, since deep-mining and old-style strip-mining have removed the easiest-to-obtain ones. Driving the finances, of course, is the price of coal. Most is sold under contract to big coal-burning electric utilities such as American Electric Power, Dominion, or Duke Energy. But a spot market for both steam and metallurgical coal still thrives, although it can fluctuate wildly. In the 1990s, when coal sold at a "freight-on-board" of $22 a ton based on the price at the railcar, the numbers worked out that eight tons of overburden could be removed to get one ton of coal and still be profitable. More recently, steam coal has been selling at up to $50 a ton, so twenty tons of spoil can be removed profitably. In the case of met coal, the ratio is even bigger, since it has been selling at $100 per metric ton at the mine and $330 a metric ton delivered to Rotterdam or Shanghai.

The first step in mountaintop removal is acquiring thousands of acres of land. This can involve buying up entire towns that may have been around since the nineteenth century. The village of Blair in West Virginia was all but destroyed in the 1990s to make way for Arch Coal's 3,100-acre Dal-Tex complex employing four hundred miners. More recently, a Massey Energy subsidiary bought up most of the Boone County, West Virginia, hamlet of Lindytown for a mountaintop-removal operation, leaving one family that took some money but decided to stay. A Massey attorney says that the firm was under no obligation to buy up most of Lindytown but did so after a number of residents approached the company. They didn't have much of a choice. Their land would be next to worthless after adjacent lots had been mined.

Three decades' worth of mountaintop-removal mining have had profound effects on the ecosystems of Central Appalachia, according to twelve researchers writing in "Mountaintop Removal Consequences" in a January 2010 edition of *Science* magazine. Among their findings were that 10 percent of the watersheds in West Virginia alone have been disturbed by the practice. Filling valleys with mine waste jeopardizes "extensive tracts of deciduous forests" that "support some of the highest biodiversity in North America, including several endangered species." Water-flow paths are mostly through forested areas in permeable soil areas, the scientists wrote, but on mined sites, the flow goes overland, not through the soil, increasing the likelihood of flooding. Waste-filled valleys are far more prone to release toxic elements, such as sulfate, calcium, magnesium, and selenium, into the environment. A survey of seventy-eight valley fill streams found that seventy-three had toxic levels of selenium, an element essential in low doses for living organisms. In high doses, however, it can cause hair loss, gastrointestinal problems, cirrhosis of the liver, and death.

Reclamation projects also have a poor record. "Many reclaimed areas show little or no growth of woody vegetation and minimal carbon storage even after 15 years," wrote the researchers. "Clearly, current attempts to regulate [mountaintop removal] are inadequate," they conclude. "Mining permits are issued despite the preponderance of scientific evidence that impacts are pervasive and irreversible and that mitigation cannot compensate for losses." Remembering when I was a boy the skeletons of dead animals next to the toxic ponds on the abandoned and much smaller strip mine sites, I find it hard to imagine the scale of destruction of mountaintop mines encompassing thousands of acres if not several square miles.

Fifty miles east of the coalfields sits the West Virginia town of Lewisburg, a quaint area with redbrick houses and the occasional cobblestone street. As such, it is more reminiscent of the prosperous suburbs of Washington, D.C. The surrounding countryside also looks like Virginia's rich horse country, with its rolling hills and picturesque farms, rather than the tight valleys, harsh hollows, and Dollar General stores of the coalfields. Not far away is the posh and historic resort of the Greenbrier, which was owned for years by the coal-hauling Chesapeake and Ohio Railway. More recently, the resort had to convert into a gambling casino, but it still retains the ambience of its grand old days with the Southern plantation-style white columns of the main building's entrance.

Here in Lewisburg, on the second floor of an old Washington Street building, is the office of the legal scourge of the coal firms involved in mountaintop-removal mining. Joe Lovett of the Appalachian Mountain Advocates is a trim young man in his forties. On one wall is a black-and-white photo of jazz trumpeter Miles Davis. A jazz fanatic, Lovett has hundreds of albums of leading musicians, including a 150-album collection of Davis. Along with the image of Davis, a blown-up news photo of the Widow Combs being carried off by Kentucky cops with flattop haircuts is behind his desk. Lovett is from West Virginia, went to law school at the University of Pennsylvania, but returned home, where he thought he might do more good and be close to his parents.

In the process, Lovett has tossed plenty of legal monkey wrenches into the regulatory wheels that have for the past decade favored the coal operators. Both mine safety and surface regulation "are failures as regulatory issues," he says. "Coal is more important than people or the environment."

Applying federal laws on permitting mountaintop-removal sites has been an arm-wrestling match for years. Oversight is split up among a number of federal agencies with different attitudes, interests, and weakness to political manipulation. The U.S. Environmental Protection Agency, which Lovett claims has been more approachable under Obama, oversees a Section 402 permit under the National Pollution Discharge Elimination System (NPDES) covering what liquids mining firms can put directly into a stream through a pipeline. Another situation involves water pollution that goes into a watershed in a more indirect way. The EPA oversees but delegates authority to state regulators to check the quality of polluted water that goes from a mine to a control pond and then to a stream. The state regulators, such as the West Virginia Department of Environmental Protection, routinely shared their reports with the EPA, but President George W. Bush, bowing to coal-industry pressure, put a stop to that practice. Consequently, the EPA often did not know what state regulators had approved.

The most nettlesome permit for Lovett is Section 404 of the Clean Water Act that is overseen by the U.S. Army Corps of Engineers. The corps is a formidable institution tasked since the Revolutionary War with maintaining the nation's waterways and dams for national defense. Some of the brightest minds at West Point end up in the corps, giving it a special panache within the military. The elitism also means that the corps, which apparently doesn't like the mundane job of dealing with coal mines, is "a very arrogant bunch" that "doesn't like people challenging them," says Lovett. The corps' Huntington, West Virginia, office has blunted inquiries from environmental groups and ordinary citizens for years.

Its oversight of Section 404, however, is critical to the operation of a mountaintop-removal project. It oversees the introduction of solid, as opposed to liquid, pollutants being introduced into streams. Mountaintop mining cannot proceed unless operators have permission to dump thousands of tons of dirt and rock waste into stream and creek valleys. Scientists have found that this method of disposing of solid mine waste chokes off needed streams and pollutes them with a toxic stew. Yet the corps has regularly overridden EPA objections and granted permits, notably to Arch Coal's controversial Spruce No. 1 Mine in Logan County, West Virginia.

Lovett has helped shoot down a program known as Nationwide 21, which evolved during the Bush administration to fast-track permitting under surface coal mines if the environmental impacts were deemed minimal. The industry had cheered the practice, claiming that overly extensive reviews were hurting the American economy and costing them money. In some cases, the fast-track permitting system meant that the public wasn't even informed that permits were granted, cutting off vital debate about a surface mine's impact. Lovett's biggest victory came in January 2011, when the EPA, evoking its authority over water-pollution law, vetoed a water permit that the Corps of Engineers had issued to Arch Coal's Spruce No. 1 Mine. If the decision withstands legal challenges, it will be the first time that regulators have shut down a mountaintop-removal project. In this case, environmental groups say, more than seven miles of streams and two thousand mountain acres will be saved.

Lovett, however, isn't crowing about the victories. The economics still favor mountaintop removal, so "they're going after what's minable today," he says. One problem is that the ravages in the coalfields are still far off the radar screens of the rich and influential. "This isn't Vermont. The rich don't vacation here," he says. Nor is he especially hopeful that Alpha Natural Resources will do a better job than Massey Energy did with the environment after it takes over the firm. "Why would they?" he asks. Lovett has a point. Alpha is under the same economic pressures as Massey to keep its costs down, and as natural gas challenges coal in the electricity market, cost cutting will become more important to keep prices down. This trend would tend to encourage coal firms to dodge regulations and avoid steps to preserve the environment.

For Vernon Haltom, the cycle of protests and publicity continues. But despite the demise of Massey Energy, he has to keep his head down more than before. On June 23, 2009,

Haltom was helping lead a protest against Massey and mountaintop removal at the Marsh Fork School just a few miles
from Upper Big Branch. Attending were Daryl Hannah, the
Hollywood actress, and about two hundred others. Confronting them were a number of family members and friends of
miners working for Massey. The most aggressive were miners'
wives, Haltom says. Hannah was arrested along with NASA
climatologist James Hansen. Activist Judy Bonds was slapped
about the head, ear, and jaw by a Massey supporter, says Haltom. Another woman yelled at Haltom's wife, Sarah, "We
know where you live. Watch out!" Because of the threat, Vernon and Sarah decided it would be safer if they moved away,
so they ended up in Princeton and then Athens, West Virginia, about forty-five miles to the south. It is near Haltom's
teaching job and gives him and his wife peace to raise their
new baby.

He still commutes to Whitesville, and the protests continue. A few months after the slapping incident at Marsh Fork,
in October 2009, environmental activists tried to enter a civic
center in Charleston where the Army Corps of Engineers was
holding a hearing on the Nationwide 21 program. It was a
similar drill, says Haltom. Pro-Massey people crowded the
room as officials fretted about fire laws. An eighteen-wheeler
blocked a parking lot. The driver kept blaring its horn. Inside,
"Friends of Coal" agitators tore up paper flyers, threw them
on the floor, and ordered the environmental activists to pick
them up. Tensions rose. Environmentalists demanded that

the police let them in the hearing. Finally, the police summoned Haltom and asked him to leave. "I didn't see much choice. But we did get the Nationwide 21 process stopped," he said.

It would be a small victory in an ongoing war.

7

DARK AS A DUNGEON

Gary L. Stover, a short man sporting a mustache, sits in his map-filled office in Chesapeake, West Virginia, a stone's throw from the coal-laden barges being pushed up and down the Kanawha River a few miles east of Charleston. Stover has worked in the coal business for thirty years, including stints with mining firms Consol, Peabody, and Massey Energy. His current company is Penn Virginia Resource Partners, which buys and sells coal- and timberlands in the Appalachians. Stover left Massey fifteen years before, but one event there sticks in his mind. It cost him his job and nearly his conscience.

In mid-1995, he was employed by Massey Energy's Elk Run Mine just a few miles up a twisting country road from Massey's Upper Big Branch facility. At the time, Stover regarded Elk Run as a well-run operation with a good safety record with "only a couple of fatalities." One day in mid-1995, the mining engineer with a degree from Virginia Tech was in a meeting with Don Blankenship, then the new chief executive and chairman of Massey Energy.

Stover had been flinching at the abrupt change in management style after Blankenship moved to the top. Previously, the chief had been E. Morgan Massey. Although he had an iron fist when it came to organized labor, Massey ran the business in a loose, decentralized way. While he was available if needed and operating employees could always ask him for advice, they were encouraged to make decisions based on their judgment. "There was no problem going to the president of the operation and getting an answer. It was no problem, they would consult with Morgan." Not so with Blankenship. Stover characterized him as so detail obsessed that he demanded to be consulted on even the smallest matters.

At one meeting, Blankenship leaned his stocky frame over a map of the Elk Run Mine. As is typical for an underground mine, it was arrayed in a series of underground rooms and passageways held up by earthen pillars. The lives of the miners depend upon the stability of the pillars and the integrity of the roofs they held up. As with many engineering projects, pillars at Elk Run were erected with a safety factor of 1.5, meaning that they would have enough strength to withstand the weight of the earth above them then one-half times that weight. In another example, a bridge might have enough strength to withstand weight, stress, wind, and other factors times one plus one half or perhaps more for safety. "It was a pretty widespread industry standard," says Stover.

Blankenship tapped his finger on a map of a pillar. "What's the safety factor here?" he asked. "One point five," I said. Blankenship replied, "We'll have no more pillars with a safety factor of more than one," according to Stover, who was stunned by the order. He believes that Blankenship's reasoning was that Elk Run would last only another five to ten years before its coal reserves would be gone. Lowering the safety factor could make pillars thinner and allow for better productivity and higher profits. But, says Stover, "It was against everything I was taught as an engineer."

Stover balked at the order, refusing to sign documents authorizing it, which led to his eventual firing. "I was demoted from chief engineer to mine engineer. Another man was promoted to my job who wanted me to certify the 1.0 safety factor. I refused. Within two or three months, we had a fatality," he says. The Mine Safety and Health Administration later forced Massey to go back to the 1.5 safety standard. Regardless, Stover's fate was sealed. He was gone by May 1997.

Stover's story is just one of many illustrating how Massey Energy's penchant for cutting corners and squeezing costs earned it the most notorious modern record in mine safety among all U.S. coal firms. Margins for error are extremely slim—something I realized firsthand when I visited Massey Energy's Red Ash Mine in Southwest Virginia. After a short safety briefing and a fitting out in a self-rescuer oxygen tank, battery pack, safety boots, and jacket, I squeezed into a man-trip that took me nearly five miles and several hundred feet down into a mountain. We were in low coal, making me bend over as I fought claustrophobia. My helmet, emblazoned with decals citing JOY mining equipment and a viper baring its fangs, constantly smacked against the ceiling of rock. Mining machines with ultra-hard bits spitting sparks ripped out coal with a banshee scream as water sprayed to keep dust down. The electric cables powering the machine whipped about like snakes. As miners, monk-like, moved amid the din speaking no more words than necessary, shuttle cars moved in to scoop up the coal and haul it to a nearby conveyor belt that whisked it to the surface. Adjacent to conveyor belts were marked corridors that we had been briefed to use as pathways to safety in case of an accident.

Any slipup or loss of concentration could be deadly. The low ceilings and limited space made it hard for miners operating powerful equipment weighing scores of tons to see other miners who could be easily crushed to death. Tangles of electrical cables threatened death by electrocution. Workspaces are so limited that escaping a fire, explosion, or deadly gas would be extremely difficult. Yet coal miners everywhere endure similar threats, making it all the more remarkable that Massey Energy was able to become king of the hill when it came to safety violations. It had the most safety violations from state and federal regulators of any coal company in the nation. Several of its mines had double the average number of safety incidents of any of its deep-mine competitors, including Patriot Coal, Consol, Arch, Peabody, and Alpha Natural

Resources. Massey has been accused of any number of oversight lapses, such as failing to ventilate its shafts properly, allowing too much explosive coal dust to build up, operating defective and shoddily maintained equipment, and increasing the risk of accidents by reducing the number of miners on a shift. As in the Stover case, if employees spoke out, they feared firing, a strong incentive to stay quiet.

When Alpha Natural Resources eventually bought Massey in June 2011, analysts and others in the industry believed that that firm with its better safety record might change Massey's corporate culture and retrain its workers through its own internal program called "Running Right," and Alpha has put six thousand former Massey workers through the program. Critics, however, note that the slogan that it touts sounds curiously similar to Massey's "Doing the Right Thing." Observers have scratched their heads that a number of top Massey executives who were involved in disasters such as Upper Big Branch have ended up in top jobs co-planning safety at Alpha. That firm's chief executive officer, Kevin Crutchfield, told me that Alpha had hired several Massey managers who had taken the Fifth Amendment to protect themselves against self-incrimination when they were asked to be interviewed by state and federal regulators. "We thought about it a long time and decided that they had a right to take the Fifth," he says.

Others are more skeptical. "I don't know if Running Right is the same as Doing the Right Thing. We'll see," says Bruce Stanley, a lawyer with the Pittsburgh office of Reed Smith who has sued Massey on several occasions and grew up in the same region as Don Blankenship. Critics point out that Massey's historic loathing of organized labor contributed to its mine deaths and injuries. Although Alpha is roughly 30 percent unionized, most of its organized areas are in Pennsylvania and Northern West Virginia, far from the mines it acquired from Massey. And, like Massey, Alpha sends its miners official "reminders" that they don't need to join a union to enjoy job security.

Massey Energy officials have steadfastly insisted that they have always paid strict attention to safety guidelines. Days after the Upper Big Branch disaster, Blankenship told reporters, "I know and am comfortable as a manager that we do everything we can to meet the law and go beyond the law." Statistics and company strategy show otherwise. After Blankenship took over the leadership of the firm in 1992, Massey adopted an approach that was unusually tough for a mining firm, both in terms of how it dealt with internal safety concerns and how it handled outside regulators and the general public.

Examples abound of Massey ignoring, or even punishing, employees when they made complaints about safety at their mines. According to an independent probe of the Upper Big Branch incident ordered by the former West Virginia governor Joe Manchin and led by J. Davitt McAteer, a former Ralph Nader activist, union lawyer, and government regulator, Massey had a hard fist when it came to safety concerns.

Institutional secrecy common among intelligence agencies was introduced, in which workers had to have a "need to know" about safety problems at mines such as Upper Big Branch, and "call out" production reports that were required every thirty minutes during a typical shift kept attention focused on productivity and made workers loath to act to shut down production because of an unexpectedly dangerous mining condition or faulty equipment.

To keep workers in line at nonunion Upper Big Branch and other company facilities, Massey used one-sided employment contracts called "Enhanced Employment Agreements." With these, Massey employees agreed to a three-year contract in exchange for guaranteed work with pay increases and bonuses. But they also were discouraged from questioning management on issues such as safety or they could be fired for offenses such as "lack of performance" or a "serious safety infraction." If this happened, they were stuck with a severe "non-compete" clause that forbade them from working at a competitor's coal mine within a ninety-mile radius of the Massey facility. This would effectively end their mining careers. Plus, they had to return all the bonuses or "enhanced pay," including tax, Social Security, and Medicare withholdings, for the period. Fired miners often had to dip into their personal savings to square themselves with Massey if they were canned. For those remaining on the job, the point was clear.

The hardball approach extended to dealing with regulators. Safety citations were quickly and aggressively challenged in the bureaucracy or in court. Massey executives were willing to eat the legal expenses, knowing that their chances were good that they might beat down state or federal safety regulators, who were tasked with policing plenty of other mines and might tire out fighting Massey on so many fronts. According to the May, 2011 McAteer report: "Massey's Vice President for Safety Elizabeth Chamberlin reportedly took a violation written by an inspector, looked at her people and said, 'Don't worry. We'll litigate it away.'" At Upper Big Branch, Massey delayed or contested 85 percent of the safety violations leveled against it in 2009. At a press conference on December 6, 2011, where MSHA's final report on the disaster was unveiled, MSHA assistant secretary Joe Main openly bristled when asked if Massey's pugnacity in contesting regulatory citations intimidated his agency. He said no, but it clearly gummed up the regulatory process.

Blankenship underlined his tough-guy persona by launching an extensive and personal campaign against "overregulation" by government agencies and meddling by labor unions—a throwback to the 1985 strike when he ducked bullets. While he became a monster to environmentalists and unions, his acts were quietly cheered on by many in the nation's conservative business community because Blankenship was willing to state openly what many of them were too afraid or circumspect to say. That said, those cheers didn't necessarily extend to his own industry. Some executives at other coal firms muttered that he was taking things too far and was drawing too much negative attention from the media, regulators, and investment community to coal firms with better safety records than Massey Energy had. They had enough problems as it was with Big Coal's traditionally dirty image.

Not even Blankenship, however, can argue with Massey's documented and dismal safety record. Over the past two decades, the firm has built a record that critics say shows a consistent disregard for safety. By the time of the Upper Big Branch disaster, the firm or its subsidiaries had four mines—Tiller No. 1 (Knox Creek Coal Co.), Slip Ridge Cedar Grove (Marfork Coal Co.), No. 1 (M3 Energy Mining Co.), and Mine 1 (Solid Energy Mining Co.)—in West Virginia, Kentucky, and Virginia that had double the average injury rate of 4.03 injuries per 200,000 worker hours.

Upper Big Branch's injury rate was 5.81 in April 2010. Judging by the number of Mine Safety and Health Administration citations, it was clearly a disaster waiting to happen. In 2009, Upper Big Branch had forty-eight "unwarrantable failure orders," citations issued after MSHA determines that miners have not corrected serious safety violations. In one case, it was fined $66,142 for allowing combustible materials to build up in workspaces. Taken together, Upper Big Branch had 515 citations for fines worth $897,325 that year. Only one other Massey mine, a Ruby Energy facility in Mingo County, had more fines—$1,668,408—in 2009. With a well-managed company, neither mine would ever reach such levels of fines.

The McAteer report concluded:

> The company broke faith with its workers by frequently and knowingly violating the law and blatantly disregarding known safety practices while creating a public perception that its operation exceeded industry safety standards. The story of Upper Big Branch is a cautionary tale of hubris. A company that was a towering presence in the Appalachian coalfields operated its mines in a profoundly reckless manner, and 29 coal miners paid with their lives for the corporate risk-taking. The April 5, 2010 explosion was not something that happened out of the blue, an event that could not have been anticipated or prevented. It was, to the contrary, a completely predictable result for a company that ignored basic safety standards and put too much faith in its own mythology.

Coal mining has been one of the most hazardous endeavors since some of the earliest mines opened in Midlothian, Virginia, in 1730. By the early nineteenth century, colliers were active in England, picking up strength with the dawn of the Industrial Revolution and its dependence upon steelmaking and, later, electricity. As demand for coal increased, so did the risks for miners, leading to some horrific disasters. In the late 1800s, a series of accidents killed more than 1,200 mine workers in the United Kingdom. Death kept pace with the demand for coal as the Steel Age expanded. On March 10, 1906, 1,099 French miners were killed at an explosion at the Courrières Mine. In China, on April 26, 1942, 1,549 miners lost their lives at the Benxihu Colliery. The worst disaster in the United States was at Monongah, in the northern part of West Virginia, on December 6, 1907, which killed 358 people. That was the deadliest year ever in U.S. mining history, a year that claimed 3,242 dead for coal and noncoal mining. Recent U.S. death rates are tiny compared with those in China, the world's leading coal producer, which mines roughly three times as much as the United States does. In 2009, some 2,631 coal miners died in China. In 2010, the death toll was 2,433.

Coal's deadly image hasn't really held up in recent decades, at least not in advanced industrialized nations. Since 1970, according to the National Mining Association, coal production has increased 76 percent while fatal injuries have decreased 93 percent. The progress results from better mining techniques and safety gear but also because far fewer miners can mine more coal. Coal-industry employment has plummeted from 749,000 in 1925 to a little more than 100,000 now. As mining became more mechanized and far fewer miners were exposed to dangers, coal dropped statistically in its ranking of most hazardous jobs. Recent data listing deaths and injuries per 100,000 workers shows that while still hazardous, coal mining is a safer line of work than commercial fishing, the number one killer of workers. Following it are logging, farming, construction, and sanitation. Airline pilots have died more frequently in accidents than coal miners in recent years. Total mining deaths dropped from 426 for the years 1966–70 to 62 for 2001–05.

So, why, then, is Massey Energy's safety record so poor when the overall trends are for safer mines? Why couldn't the company direct some of its rich earnings back into better mining equipment, safety gear, and working conditions? One reason, suggested by investor lawsuits, is that Massey was struggling to keep up with Asia's intense demand and was cutting corners to meet it while boosting its stock prices, which had slipped significantly before the Upper Big Branch disaster. And Blankenship did this by following E. Morgan Massey's Massey Doctrine, the one that advocated the ethos of the scrapper upstart who uses leftover gear and is severely anti-union. Massey regularly deployed nonunion contract labor whenever practicable to avoid paying high wages and benefits and to shift legal blame if an accident occured. Worker welfare appeared low on Blankenship's list of priorities.

As is often the case in the coal business, the never-ending cycle of disaster and tougher regulation changed his approach, at least for a while. One year after the 1968 Farmington disaster, sweeping new mine safety regulations called for by the Federal Coal Mine Health and Safety Act were approved. MSHA itself was created by an administrative act of the U.S.

Interior Department to act as its independent regulatory arm and avoid conflicts with Interior's other tasks of developing mineral reserves. And A. T. Massey, Massey Energy's predecessor, was forced to change its ways. To do so, it had to take itself public to acquire the capital for major upgrades of its mining equipment to meet the new rules.

While E. Morgan Massey was a frugal businessman and had a reputation for hardball negotiating that included breaking the back of the United Mine Workers of America in the mid-1980s, he never shared Blankenship's distaste for safety regulations. Yet much of the old philosophy prevailed and, in fact, was taken into more rigorous and tougher directions.

By the first decade of the twenty-first century, the relative quiet that had prevailed around the Appalachian coalfields was about to be shaken up. In 2001, a disaster at an Alabama mine killed thirteen, the worst since Farmington in 1968. That year also brought the inauguration of President George W. Bush and new changes to mining. His business-friendly policies included pulling back on what was perceived to be overly aggressive regulation of businesses. On the mountaintop-removal front, new, more permissive interpretations of federal regulations came along, allowing more mining spoil to be placed in streams. Bush's picks for key regulator jobs were also suspect. Named to head MSHA was David Lauriski, a former executive and lobbyist with Utah-based Interwest Mining Company, who had been chief safety officer at Emery Mining in Wilberg, Utah, in 1984, when a blast killed twenty-seven coal miners. In Washington, he drew criticism for scaling back federal rules on maintaining two or more escape ways from deep mines. Another controversial Bush appointee was J. Steven Griles, a former mining lobbyist who, according to *The New York Times*, October 26, 2006, "devoted four years to rolling back mine regulations." Griles, the number two official at the Interior Department, was cited for conflicts of interest for continuing to take $1 million in payments from his former mine company employer while at Interior. He later pleaded guilty to obstruction of justice charges involving a casino sought by convicted lobbyist and investor Jack Abramoff. Griles was fined thirty thousand dollars and sentenced to ten months in prison.

Statistics soon reflected the go-easy policies of the Bush administration. Thanks to the regulatory pullback, from 2000 to 2005 the number of citations issued by MSHA started to drop. Not surprisingly, accidents in under-inspected mines were waiting to happen. Two nearly back-to-back incidents in January 2006 brought renewed focus on mining dangers and lax regulation. On January 2, a blast at the Sago Mine near Buckhannon, West Virginia, trapped thirteen miners in the facility owned by a subsidiary of International Coal Group that had bought the mine two years before at a bankruptcy auction. During the previous year, the mine had been cited 208 times by MSHA, which was a substantial increase from the year before. For two days, families and friends waited as rescuers had to deal with high levels of carbon monoxide. Cheers erupted when it was reported that all thirteen had been reached. The report was erroneous. News media outlets were forced to correct themselves. Twelve were found dead. One badly injured miner survived. Some had managed to enter a "safe room" equipped with food and oxygen, but the miners had not been able to build a barricade from contaminated shafts. Not airtight, the barricade let poisonous gases waft into their refuge, where they ultimately died.

Fifteen days later, a conveyor belt caught fire at Massey Energy's Aracoma Alma Mine No. 1 in Logan County, West Virginia. Smoke wafted into an air intake passage that was meant to be an escape route, and two miners were killed. MSHA said that Massey failed to install needed sprinkler systems and maintain a water supply adequate to handle such underground fires. It was something that Blankenship knew about at least six days before the blaze. According to the *Charleston Gazette,* on April 17, 2010 Blankenship had sent Massey Energy troubleshooter Linton Stump to check into problems at the mine. Stump sent Blankenship a memo warning him of the mine's dangers. He alerted Blankenship that "while safety reports from Aracoma managers showed that everything was okay, indeed, it was not." Massey's subsidiary later pleaded guilty to ten federal charges, and the firm paid a $2.5 million criminal fine. Another 1,300 citations were filed in the Aracoma case, resulting in $1.7 million in additional fines.

Aracoma and Sago sparked yet another tightening of safety rules, at least on paper. New federal legislation under the 2006 Mine Improvement and New Emergency Response (MINER) Act required each mine to coordinate emergencies with local agencies and develop tracking systems to locate trapped miners and boosted required oxygen supplies to two hours for underground workers. The law was hailed as the first major reform of mine safety laws since 1977. But as is typical in U.S. mining history, the law wasn't successfully implemented, as Upper Big Branch shows.

The new law required all mines to install an electronic tracking system to show the approximate locations of miners. Such a system was being installed at Upper Big Branch at the time of the blast, but the McAteer report quotes Derrick Kiblinger, the man in charge of the installation, as saying, "The tracking system was really far behind. Maybe 20 percent of it would have been done by April 5 [2010, the date of the disaster]." He said he had trouble getting parts and needed a larger workforce to install the system. "When I started in October [2009], somebody should have been on every shift to even be close. Had this system been in place, you would have known a lot quicker where these men were. You would have known within two thousand feet, probably a little better, where they were," he said. And what happened with the new law's requirement that miners have a two-hour supply of oxygen? At Upper Big Branch, nineteen of the twenty-nine dead miners died of suffocation either because they didn't have time to don their individual air masks or their tanks ran out.

Ultimately, not implementing the latest legal safety requirements was only a small part of what led to the Upper Big Branch disaster. The tragedy was ignited, literally, when a small ball of burning gas flared up unexpectedly and brought any number of maintenance, safety training, and mine-engineering design deficiencies immediately to a head. Defending themselves, Massey Energy officials insist that somehow cracks opened up at the floor of the mine shaft near the coalface and natural gas leaked in, causing the explosion.

Among Upper Big Branch's many problems was its ventilation system. The proper flow of air into the labyrinthine passageways miles into the mountains is critical so miners don't suffocate and explosive mixtures of gases and coal dust don't build up. "Extremely low airflow was a chronic problem in some parts of the mine," the McAteer report says. Miners and foremen routinely closed air-lock doors or hanging curtains to get more air. Plus, there was evidence of major ventilation changes being made while miners were working. This, the report says, is "a blatant disregard for worker safety and a violation of law."

A basic mistake, investigators from both MSHA and the McAteer probe charge, is that Massey Energy allowed unacceptably large amounts of coal dust to build up. Dust can be kept down by dusting shafts with powdered limestone, but Massey used a twenty-year-old machine that was constantly breaking down. At a December 2011 press conference, Joe Main stated that MSHA investigators found that 90.5 percent of the 1,353 rock-dust samples they took in the mine were inadequate.

Adequate water spraying is also a common mining practice to prevent flames from touching off explosions. This, too, was problematic at Upper Big Branch. Some of the water was pumped in from the nearby Coal River, but the river water was filled with sediment that tended to clog filters on spraying devices. "Investigators heard testimony and examined physical evidence indicating that the screen and sock filters were frequently plugged so much so that the water flow to the machinery was reduced. On the Upper Big Branch longwall in particular, the river water was not filtered adequately, sediment reached the sprays, lodged directly in spray points and clogged them," the McAteer report said. "If all the water sprays had been properly maintained and had been functioning as intended—creating a fine mist of water at the shearer nozzle point—and if rock dust had been properly applied, any ignition of methane that occurred likely would have been extinguished at its source."

The report concludes: "Ultimately, 29 miners lost their lives at Upper Big Branch because these safety systems failed in a major way. Massey Energy failed to maintain an adequate ventilation system at UBB. The company failed to maintain its equipment. It failed to properly rock dust the mine. If those basic matters of safety are effectively practiced, there is no reason for miners to die as a result of explosions in 21st Century America."

One might wonder if Massey Energy's culture, which carried an almost irrational hatred of organized labor, was a flaw that otherwise could have prevented the Upper Big Branch disaster. As former Massey engineer Stover says, "If you are non-union, you are sort of policing yourself." In fact, union mines have contracts requiring safety committees that offer another level of oversight to make sure that equipment is kept running well, that coal dust is kept within regulations, that mine shafts are well ventilated, and that enough miners are on hand and trained well enough to react quickly if something goes wrong. Stover said, "I dealt with it a lot when I was face boss. And I used to tell them the same thing. If you see something unsafe, speak up. Deal with it."

Indeed, things might have been different at Upper Big Branch had it been a union mine, says Gary Quarles, the veteran miner who lost his son Gary Wayne Quarles, thirty-three, in the disaster. He's mystified about why the water was cut off at the longwall machine. Quarles also speculates that cost cutting might have played a factor. Massey usually had one man at either end of the longwall machine that traverses a thousand feet back and forth, extracting coal. "A union mine would have had two men at either end. One of them might have been more likely to hear the telephone ring and take the call. Tell them that something was happening farther down the longwall. But with only one man at either end, maybe they couldn't have heard with all that noise," he says.

It is hard to imagine the level of sound in a longwall operation. I was stunned by the screeching and whining of equipment as the longwall shearer ripped out coal as it trundled up and down a track more than three hundred yards long. This alone gives unions good cause to require two miners at either end of the machine.

Quarles says he knows of Massey's anti-union stance firsthand. Quarles, who started mining in 1976, held a series of jobs at such mining companies as Armco and Peabody. After an extended period of being laid off and working as a security guard, he learned of an opening at Marfork, a Massey operation. "I had a friend already working there and he prepped me for the interview—what to answer when they asked questions about unions. They asked me if I would cross a picket line. The answer is that you'd drive around it and go to the nearest telephone, call in, and ask what to do," Quarles says.

The United Mine Workers of America, smarting from its intense, years-long friction with Massey, is quick to note that nonunion mines are statistically less safe than union ones. "Every year [since 2000 and not including Upper Big Branch], like clockwork, at least one person has been killed on the property of Massey or one of its subsidiaries," union president Cecil

Roberts said at a press conference in Charleston three days after the Upper Big Branch blast. "No other coal operator even comes close to that fatality rate during that time frame," he said.

The UMWA had long targeted Upper Big Branch but failed to organize it on three separate occasions. According to the UMWA president, Cecil Roberts, one vote to certify the union as the official bargaining agent ended in a tie, which gave management a win. The union lost a second attempt in straight votes, and in a third attempt the union decided to call it quits before an election. "Had this been a UMWA mine, a miner could have shut it down by noting an unsafe condition," Roberts told a press conference at the Embassy Suites Hotel in downtown Charleston in October 2011. Roberts dubbed Massey management of the mine "industrial homicide."

What makes Upper Big Branch different from other Massey Energy mishaps is that, thanks to Blankenship's controversial persona and the sheer size of the tragedy, the incident has stirred a criminal probe that could go far beyond other criminal charges the company has faced in other disasters, such as the Aracoma fire. In an unusual step, MSHA was slow to release transcripts of miners and rescuers its officials had interviewed. Officials said they did so at the request of the U.S. Attorney's Office for the Southern District of West Virginia, which has its headquarters in Charleston and branch offices in Beckley and other towns. By mid-2011, 266 individuals had been interviewed. Eighteen executives from Massey Energy or its subsidiary that ran Upper Big Branch, including Donald Blankenship, declined to be interviewed by federal investigators, citing their Fifth Amendment right to avoid self-incrimination.

Although prosecutors made no public statements for months after the disaster, it seems that they are targeting the practice of falsifying safety and other documents. If so, they are going far beyond rapping the knuckles of a coal operator for cutting corners on safety regulations. The first clue came in February 2011 when Hughie Elbert Stover, sixty, was indicted on two felonies relating to his job as Massey's chief of security at Upper Big Branch. The federal indictment alleges that Stover (no relation to Gary Stover) had told federal investigators in January 2011 that Massey's subsidiary operating Upper Big Branch had a ten-year-old policy forbidding security guards from tipping off workers that safety inspectors were coming into the mine. The indictment says that Stover "had himself directed and trained security guards . . . to give advance notice by announcing the presence of an MSHA inspector" over the mine radio. He is also accused of telling another individual to dispose of mine security documents in a trash compactor. He was convicted in October 2011.

In late June 2011, MSHA officials told family members of the victims in a private meeting that they had found evidence that Massey had been keeping two sets of production reports. According to MSHA coal administrator Kevin Stricklin, keeping two sets of books is not in itself a crime, "but they are required to record the hazards in the official sets of books." MSHA investigators claim that the official safety records required by law did not show some of the same safety problems that Massey's internal production reports showed. One report stated the mine had problems with ventilation and may be accumulating too much methane gas. Another doctored report on the same day declared no such problems.

The MSHA report echoes many of the accusations made in the McAteer report, such as stating that faulty and improperly maintained water sprays on the longwall shearer device could have put out the small flame at the coalface and prevented the explosion. MSHA also says that a gas detector that could have registered spikes in methane gas hadn't been turned on for two weeks before the incident. Despite pronouncements by an internal Massey report and from Bobby Ray Inman, the former chairman of the firm, MSHA says that a crack in the shaft floor was too shallow to let methane waft up from beneath the mine.

The big question looming now is how and if Alpha Natural Resources, which took over Massey Energy on June 1, 2011, in an $8.5 billion buyout, will change Massey's corporate culture and safety record.

Alpha CEO Kevin Crutchfield says that his firm will retrain Massey workers. They have been introduced to the company's Running Right program, which emphasizes safety and encourages mine workers to alert management of problems even if it means shutting down production until they can be addressed. "We are absolutely committed to running right," he said. One question, however, is how this strategy can be achieved, since a number of former top Massey managers have transferred to Alpha. The man helping implement Running Right for Massey workers had been Chris Adkins, who had been Massey's chief operating officer and was in charge during the years when Massey built up its remarkable record of safety violations. Upper Big Branch had more MSHA closure orders than any other mine in the United States, and the firm's Freedom Mine No. 1 in Pike County, Kentucky, had so many safety violations that the U.S. Labor Department tried to have it placed under the supervision of a federal judge. Adkins, meanwhile, has pleaded the Fifth Amendment rather than be interviewed about the Upper Big Branch explosion.

While Alpha enjoys a reputation for being more transparent and far better at public relations than Blankenship or other Massey officials, Alpha's behavior has mirrored that of Massey. In 2011, for instance, Alpha joined the West Virginia Coal Association, an industry lobby, in opposing two sets of new safety rules proposed after Upper Big Branch. The rules would require miners to fix serious regulatory violations while they also are conducting mandatory safety checks for hazardous conditions, such as high levels of methane gas. John Gallick, Alpha's vice president of safety, says the rules would place unrealistic burdens on miners.

In the end, according to former Massey engineer Gary

Stover, common sense should prevail. Noting criticism that Massey miners badly maintained the bits on its shearers at the Upper Big Branch longwall machine, contributing to the sparks that ignited a gas flame, "The first thing you do in the morning is check your bits," Stover says. "The ultimate person responsible for safety is you. I have a daughter and her boyfriend is a miner and I tell him that all the time. The guy you have to answer to is the guy you look at in the mirror every morning."

Yet self-reliance can go only so far when working in tiny spaces thousands of feet down and miles into a mountain. Lives depend on the competence of coworkers plus dependable equipment and a professional management. At Upper Big Branch, there is ample evidence that the latter two items were in short supply or nonexistent, thanks to Massey Energy's mania for productivity and putting the squeeze on its workers. Perhaps Alpha Natural Resources, Massey's new owner, will do better given its superior safety record. But hiring the same Massey managers at similar positions overseeing safety and its anti-union messages dampen hopes. And it will take years for the families of Central Appalachia to heal from the loss of their loved ones.

8

COAL COUNTRY CULTURE WARS

Don Blankenship's attire was as preposterous as his message. Dressed in a U.S. flag–accented red, white, and blue polo shirt and topped with a dark-blue sports cap sprinkled with stars, he stood on a stage before a crowd of seventy thousand, milling atop a bulldozed-over mountaintop-removal strip mine near Holden, about ten miles west of Logan, West Virginia. On deck for the Labor Day 2009 extravaganza were such popular entertainment luminaries as right-wing radio jockey Sean Hannity and musicians Hank Williams Jr. and Ted Nugent, the aging rock star and conservative lightning rod.

Titled Friends of America, the rally would be one of the last public venues for Blankenship, who was enjoying the large and friendly crowd. Another treat was that his Labor Day celebration competed directly with another rally sponsored not far away by the United Mine Workers of America.

"It was Don's way of rubbing it in to the union since it was

Labor Day," says Pittsburgh lawyer Bruce Stanley. "He's having the festival atop a mountaintop-removal mine site, so he's sending the appropriate message to the environmentalists. I don't know if Don has a politically correct bone in his body."

Boorish and emboldened, Blankenship had cleverly sensed a critical fissure in his war against what he sees as the political forces that threaten his coal industry along with the United States. His rally would exploit Appalachian culture as a way to further split the divides between pro-coal and its union and environmental critics. The idea was to polarize haves and have-nots by presenting criticism of coal as a threat to livelihoods. His strategy is much the same as the Tea Party movement identifying the downtrodden "average" people and making bogeymen out of big government and the cultural elite that live on either coast. Even with Blankenship's retirement, it's a schism that continues on to this day.

No other modern chief executive in the United States had gotten so down and dirty in the country's cultural wars. Don Blankenship relished his chance to cast himself as some kind of neopopulist Huey Long championing what he considered "his" people in ways that exploited conflicts pitting rural versus urban, blue-collar versus white-, and poor versus rich. For all his fervent antiunionism, Blankenship was paying a lot of money and spending a lot of time reaching out to his version of the proletariat. His audience may be based in his Appalachian homeland, but it might as well have touched every fast-food joint, truck stop, pool hall, and two-step parlor from Seattle to Savannah.

Despite the reality that Massey Energy was a tough and unforgiving company to work for, Blankenship succeeded in tapping a huge store of working-class resentment in ways that bought him political cover. In the name of jobs, patriotism, and protecting the constitutional right to own firearms, he used celebrities to incite the masses to pressure coalfield politicians to water down proposed laws that would hurt his profit margins and limit how he produced coal above- and underground. His cultural enemies were convenient straw men and women—affluent elites and their dupes who claimed to love the Appalachian coalfields, might visit on occasion, but otherwise wouldn't be caught dead moving permanently to them.

Blankenship's personality has been a lightning rod for any number of environmental groups ranging from the grassroots Coal River Valley Watch directed by Vernon Haltom to well-heeled, big-money types who contribute to or run such major ecology outfits as the Sierra Club and the National Resources Defense Council. The anti-Blankenship crowd extends into global finance, including the Rain Forest Action Network, a San Francisco–based organization that tracks the environmental record of global corporations and lobbies big banks such as JPMorgan Chase or the Bank of America not to extend loans to offending companies. Their work is paralleled by the sophisticated institutional investors such as managers of big pension funds representing teachers or state employees in California, Connecticut, and New York along with the Sisters of St. Dominic of Caldwell, a small band of Roman Catholic nuns who perform a similar mission from their New Jersey convent.

Anti-coal activists deploy their own name-grabbing forces at consciousness-raising events as well, including Daryl Hannah, American music icon Emmylou Harris, and Nashville star Kathy Mattea, whose activism is matched by impeccable credentials, since she is a West Virginia native with coal miners on both sides of her family.

At Blankenship's Labor Day event, however, Mattea and Harris seemed a million miles away. Fans were wildly applauding Blankenship's words. They had started showing up at six o'clock that humid morning. Warming them up had been emcee Nugent. The strident gun rights advocate had been a member of the Amboy Dukes, the 1960s rock group that dressed up as London dandies and brought the world the surrealistic song "Journey to the Center of the Mind." Borrowing a riff from the rising Tea Party movement, Nugent declared: "I love America, but mostly I love defiance. I like it when the punks from England overtax us and we throw the tea in the bay. Isn't that a cute move? But I particularly like it when the British came to get our guns so we went to Concord Bridge and shot them. I like dead tyrants. Isn't that your favorite type of tyrant, a dead tyrant?"

Tying his message to coal and Massey Energy, he posited that "people are . . . stupid" enough not to make the connection between having convenient electric power and mining Appalachian coal. "Quit regulating us based on what some pierced-eared hippie in San Francisco who thinks we're destroying the environment while she puts on a light switch, and the light comes on. On behalf of the Nugent family, I say, start up the bulldozers and get me some more coal, Massey."

More supporting themes include backing the coal industry against ignorant and selfish interlopers. The West Virginia–bred country trio dubbed Taylor Made and composed of singers Brian and Greg Duckworth and their sister Wendy Williams belted out such coal-inspired messages as "The West Virginia Underground." Lyrics include: "This here's coal mining country. That's what we do and we don't like you nosin' around."

The song is also a message in a message. Taylor Made member Brian Duckworth, the son of strip miner, told me he wrote the song because he was deeply annoyed that movie actress Hannah got so much publicity for getting arrested at a mountaintop-removal protest against Massey Energy near Whitesville in June 2009. Hence the stanza: "This ain't Hollywood,/It's West By God,/It's Virginia and Kentucky coalfields./It's the way of life that's feeds the kids and wife." The kicker: "Unless you're paying my bills, get your ass off my hill, we're the West Virginia Underground."

The marquee name at the rally was a veteran country

music star with a long career of ballads about the virtues of self-confidence and abuse, memories of his famous father, and rural bromides about partying with a pig in the ground and beer on ice. Hank Williams Jr. told the crowd how magical it was to grow up in a house with his dad and frequent guests such as Fats Domino, Carl Perkins, and Jerry Lee Lewis. With that he broke into "Whole Lotta Shakin' Goin' On." Then, milking his bestselling ode to rural self-reliance, "A Country Boy Can Survive," he took a swipe at environmental activists, shouting out: "We make our own whiskey and our own smoke, too, a lot of things our *beautiful people* can't do."

Of all the individuals who have faced down Blankenship and Massey Energy in the public arena, no one fits the "beautiful people" definition better than Robert F. Kennedy Jr., one of the best-known members of the fourth generation of the famed political clan. One hot afternoon two years later in June 2011, Kennedy, now in his late fifties, showed up at a freshly cut grassy area in the tiny town of Blair, West Virginia, not all that far from where Blankenship held his America-fest eighteen months earlier. Before a crowd of about one thousand, he was dressed modestly in pressed khaki pants and a blue and white checkered shirt. Kennedy, who suffers from spasmodic dysphonia, a medical disorder that makes it hard for him to speak publicly, paused a few moments to catch his breath and prepare for his speech.

Kennedy was wearing a red bandanna in memory of the armed coal miners who, in 1921, marched from the town of Marmet near Charleston to Blair in Logan County, which was controlled by a hard-nosed and anti-union sheriff. Thousands strong, the miners opened fire in a skirmish with deputies that quickly exploded into a full-fledged battle with rifles and armed vehicles. The Battle of Blair Mountain was put down only after U.S. troops, backed by tanks and machine-gun-firing biplanes, fought and tear-gassed the strikers into submission. The miners wore red kerchiefs and some believe that's one version of where the term "redneck" came from.

On this day, some of the demonstrators, mostly in their twenties, had finished a five-day-long hike from Marmet, tracing the steps of the miners back in 1921. Kennedy was to speak to the group, followed by others. Those capable of it then would hike up to the summit of Blair Mountain, which has been marked for mountaintop removal by Massey's new owner, Alpha Natural Resources.

Walking to a makeshift stand, it was hard not to think of his father, Robert F. Kennedy, who is said to have been more dedicated to the more radical social justice principles of Roman Catholicism, right up to the point where he was fatally shot by an assassin on June 5, 1968, when he was in Los Angeles running for president. At the time, Bobby Jr., the third of his father's eleven children, was a slightly built, blond fourteen-year-old at a boarding school in Maryland. Today, he's a professor of environmental law at New York City's Pace University and a global environmental activist.

Kennedy said that he was there because his father was radicalized by the poverty in West Virginia when he was helping his brother campaign for president in 1960. When he was a boy, he said his father told him why West Virginia is first in natural resources but forty-ninth in poverty. One reason, the senior Kennedy said, was that strip-mining was a cheap way to get at coal and put union miners out of work. RFK Jr. then launched into mountaintop removal, which he bitterly opposes. "If you came to the Hudson River and you tried to fill twenty-five feet of a Hudson River tributary, we would put you in jail. I guarantee it. If you tried to blow up a mountain in the Berkshires, the Adirondacks, or a mountain in Colorado, California, or Utah, you would go to jail." Don Blankenship "has had sixty-seven thousand violations of water pollution laws" and represents "the dominance of corporate interests on government," declared Kennedy as a sign promoting *The Last Mountain*, a documentary movie on mountaintop removal that he coproduced, flapped nearby.

The cheering crowd was an eclectic bunch—some leftover 1960s hippies and others dressed in the torn attire, tattoos, and body piercing of a 1990s grunge band. Middle-aged men with big-city accents from the Northeast wearing red "Trainer" T-shirts and Birkenstock sandals huddled on the fringes, giving intense lectures to protestors young enough to be their children. "When you go on the hike up the mountain," one thin New Yorker said, "make sure everybody is hydrated. You don't want them passing out up there and then have to carry them off the mountain." One woman from Delaware explained she was there because 89 percent of her state's electrical power comes from mountaintop removal. A pair of perky women who were twins grinned and hammed in front of a camera with a sign saying that WE LOVE OUR MINERS. One of the twins, a thirty-something wearing a rhinestone-covered cowboy hat, said they were from the Northern Virginia suburbs of Washington but had taken an interest in a West Virginia town that was about to be affected by mining. The mine, Frazier Creek, had been bought by an Indian firm, they said. Local folk needed to be educated about what was going to happen, one added.

Remembering that India has a large, fully integrated steel industry that includes blast furnaces needing coking coal, I asked, "So it's metallurgical coal?" to which one of the twins responded, "What's that?"

While there was a heavy presence of state troopers and sheriff's deputies, the event proceeded peacefully. Prior to the Blair rally, marchers had been denied camping permits and run out of several towns by police who videotaped them. At the rally, a few stray anti-activist locals occasionally got into shouting matches with the protestors, but they were quickly pulled apart by state police in dark-green uniforms. On State

Road 17, which runs through Blair, one of the few modest, white-paneled houses in Blair was decorated with signs stating A PROUD COAL MINER LIVES HERE and FRIENDS OF COAL along with a scarecrow on a front-yard stake wearing a Massey miner's jacket. Later, I watched the crowd from the road. Standing next to me was a miner missing some front teeth who apparently was returning home from a shift. I asked him for his thoughts. "Reminds me of Woodstock," he said. Neither event seemed to have impressed him.

The 1960s continually echo through coal country's cultural wars. The Kennedy name conjures up the New Frontier, when dynamic, idealistic men tried to drag the nation into confronting bitter truths about race and poverty while betting on a better tomorrow. Yet Bobby Kennedy Jr. is as flawed as his namesake. Despite his advantages, he suffered the tragic murders of his father and uncle, the unforgiving glare of publicity, two divorces, and a criminal rap involving possession of heroin in South Dakota (record now expunged). He resurrected himself from probation as an activist lawyer and chairman of the Waterkeeper Alliance, an environmental group to protect the Hudson River and other waterways, but Kennedy has also drawn criticism for using his family wealth to charter jet aircraft to fly to environmental rallies and writing error-filled articles for national publications. In early 2011, *Salon*'s editor retracted an article Kennedy had written for the Web site along with *Rolling Stone* magazine alleging that the federal government was covering up connections between childhood autism and inoculations against disease that contained mercury. Kennedy's piece had been dogged by allegations that his facts were wrong. Further investigation of Kennedy's piece "eroded any faith we had in the story's value," said *Salon* editor Kerry Lauerman, who removed the article.

Blankenship, not much older than Kennedy, all-too-neatly fits the Archie Bunker role that reacts against the Kennedy liberalism. His are the politics that resent big government with its rules and bureaucrats. It's almost as if he's a giant, endless quotation from Ronald Reagan lecturing about how the nation desperately needs to get back on the right track after the excesses and conflicts of racial integration, the Vietnam War, the drug culture, welfare queens, and the sexual revolution. The so-called left-wing news media has also stirred Blankenship's wrath. When an ABC-TV cameraman ambushed Blankenship at Massey's operational headquarters—essentially a double-wide mobile home with an electronic fence and a heliport—next to U.S. 119 in Belfry, Kentucky, Blankenship roughly shoved him away. The remarkable video went viral on the Internet and is still available. It shows a personal confrontation unimaginable to just about anyone occupying America's corporate C-suites. "He is such a great caricature," says Joe Lovett, an environmental lawyer in West Virginia who has fought Massey Energy for years.

As Pat Garland, a friend of Blankenship's who is a retired

Baptist missionary and now runs a bed-and-breakfast in Matewan, West Virginia, puts it: "Don worked several jobs at once to get through college. He slept on the floor. Can you imagine how hard he worked? Are you on his side or are you for that other guy, the young lawyer [Kennedy Jr.]?"

So it may have been preordained that the pair would meet in public debate as they did on January 21, 2010, just three months shy of the Upper Big Branch disaster and between their dueling rallies. The ninety-minute discussion was held before a sold-out crowd in downtown Charleston with the overflow ending up in an adjacent gymnasium watching on giant television screens. The arguments were predictable. Blankenship said that the "mission statement" for coal is this nation's prosperity. Kennedy retorted that coal companies were causing "the worst environmental crime that has ever happened in our history." They were "liquidating this state for cash with these gigantic machines."

The arguments may have been familiar, but the fact that a CEO of a major American corporation was willing to debate an opponent publicly in such a way says a lot about Blankenship. It suggests the complexity of his personality—that as cartoonish and boorish as he may seem, he's not afraid to sit on a stage before thousands of people without a teleprompter or platoons of public relations people and go at it with the scion of America's most famous political family, who obviously has most of the big media on his side. As contradictory as it may seem, Blankenship's fire-in-the-belly fervor can often seem admirable if misdirected.

That debate has long ago spilled over into one of Appalachia's favorite art forms—music. The Scots-Irish ancestry of many of the people of Central Appalachia has always made songs, especially ballads, a preferred way of remembering history or making a point. The tradition is as strong today as it ever was. In Appalachia, the cultural genre began as "protest" singing in the late nineteenth century. One originator was Aunt Molly Jackson, born in Clay County, Kentucky, in 1883 to a family who operated a store for coal miners. Her youth was marked by illness, poverty, and mine death. Her pro-union activities brought ostracism, and she later fled to New York City because she and her family had been blacklisted from employment by the coal operators, says Bill C. Malone, a retired history professor who taught at Tulane, Duke, and the University of North Carolina at Chapel Hill and is an expert on American folk music. Aunt Molly linked up with big-city radicals and recorded her first album protesting mine abuses in 1931. Her work was also recorded by the Library of Congress in 1940.

Another major influence was Jean Ritchie, who was born in 1922 and grew up in the coalfields of Kentucky's Cumberland Mountains. Skilled at picking the dulcimer, Ritchie didn't realize until the late 1940s that what she was playing was called "hillbilly" music until she heard it so named on the radio. After college, she also ended up in New York City doing social work and joined the nascent folk music crowd, where she became "the Mother of Folk." Her song about a faded Kentucky coal mine town, "The L&N Don't Stop Here Anymore," was chosen by Kathy Mattea for her 2008 CD *Coal*.

Mattea, who, like Ritchie, got her start in bluegrass, became famous as a country music star in the 1980s and is today's torchbearer. She grew up in Cross Lanes, West Virginia, near Charleston, the granddaughter of coal miners. Her mother worked as a secretary for the United Mine Workers of America. Mattea had a talent for singing and played in a bluegrass group during her college years in the 1970s. That morphed into mainstream country, and she signed with Mercury records in 1983. By the end of that decade, she had a number of hits such as "Eighteen Wheels and a Dozen Roses" and "Goin' Gone." She would win four Country Music Association awards, including Female Vocalist of the Year. Yet it was when she was nineteen and working as a guide at the Country Music Hall of Fame in 1978—the epicenter of the country music glitter of Nashville—that she studied films shown at the museum focusing on veteran singer Merle Travis's rendition of the coal-mining song "Dark as a Dungeon," and it was those films that quickly politicized her views on coal.

In 2006, Mattea was invited to perform at a memorial for

victims of the Sago Mine disaster. The service rekindled an awareness of coal that has been with her since she was born. She told me that she sees the current conflict as terribly polarizing and of great concern to her because she's a born and bred Mountaineer with a father from Fayette County and a mother from Putnam County. She says the conundrum with the current debate is that pro-coal activists see things only in the short term—that coal is needed now, regardless of what the long-term impacts will be. "I'm watching the water being contaminated and the mountains being destroyed in a way that nothing can be replaced."

Another tragedy in West Virginia, she says, is that coal is the only big industry with money—something Blankenship has aggressively exploited. She recalls a friend from the 1980s strike against Massey. "He held the picket line. They would bring people in from out of state. They'd drive across the picket lines and wave hundred-dollar bills. He just said they have broke us. They couldn't do it anymore."

She's pessimistic that Alpha Natural Resources, which has bought Massey, will bring improvement. "I have no way of knowing that, except that they have not changed many of the executives. I don't have much hope."

Coal companies have tapped the trend toward short-term thinking in today's cultural wars, she says. One man who tried to buck the trend, she says, is Cecil Roberts, the head of the United Mine Workers. He wrote that he took flak from his own constituents when he attended a conference focused on future alternatives to coal, whose eventual demise is inevitable. Roberts's strategy, she says, is to be reasonable so pro-environmental activists can get a seat and a voice at the table. "His viewpoint gave me hope," she told me.

As the coal trundles out of state in railcars, no one is addressing the needs of her native state. "You can measure financial stuff, but the health issues of people tucked away in the hollers of West Virginia, it's harder to bring a real quantifiable measurable study, even though you have real people experiencing it. They go from living in heaven to living in hell overnight. It's awful. It's the big shift we've had culturally in the past fifteen to twenty years." She sees her role "as trying to put my energy into how we can have a conversation." The obvious way is to continue recording music that highlights the issues, she says. "Am I just going to scream and yell, or am I going to try and do something positive? How can I use that energy so I am not just part of a movie that's been going on for fifty years, seventy-five years, or a hundred years? I am tempted to stand up and scream. This is harder."

Plenty of country music has a political point of view, from the flag waving of Toby Keith to the left-leaning Dixie Chicks as do the culture wars of the Central Appalachian coalfields. In Eastern Kentucky, coal miner Jesse Mullins struck back at the anti-mountaintop-removal activists by claiming that the area's Walmart and Lowe's big-box stores, hated in many towns but considered community assets in poor, rural areas, need stable and flat land created by reclaiming mountaintop-removal mines. His song, "Hey, Tree Hugger," mocks environmentalists and President Barack Obama. Mullins has a ready market in coal-industry lobbying groups. It has been posted on the Web site of the Virginia Mining Association, and the West Virginia Coal Association makes it available for downloads as cell phone ringtones.

The West Virginia–bred, pro-coal music group Taylor Made, who performed at Blankenship's 2009 Friends of America rally, told me that they were annoyed that the coal industry, which means much to their state, was being attacked regularly by outsiders. "Being born in West Virginia, you have to support coal," says Brian. "We make it clear we are friends of coal. We support the guy who digs the coal, the guy who buys his groceries at his grocer, the guy who buys his gasoline at the convenience store. We're all for that—for people who need to make their livings off of coal." The band picked its name because they grew up in Taylor County near the small railroad town of Grafton. They all sang in church, and their father worked at a strip mine and was a cement mixer. Brian and Wendy stayed in the northern part of the state while Greg, a state trooper, lives near Beckley. In 2004, they decided to form Taylor Made and at first the songs were apolitical. As anti-coal protests got louder, his response was the song "West

Virginia Underground," a tough-worded anthem to coal miners. "At some time in the future, coal isn't going to be necessary, but there's nothing to replace it yet," Brian says. "We only have so much coal for the next thirty or fifty years. In the meantime, we don't see any solutions from them [the environmentalists] yet. We understand that coal is on the decline, long term. We're not complete idiots."

Kathy Mattea has little use for Taylor Made. Asked about their views, she spits out that she knows their song "West Virginia Underground" very well. "I've had it blasted at me by people driving by," she told me. "Why don't they write a song about Don [Blankenship]? That would contribute something— that might crack open somebody's heart. Instead, they write a song that polarizes everybody and promotes the status quo, and they make a whole bunch of money from the people who want to go buy it and blast it."

Professor Malone echoes Mattea's point. He says that the coal industry has an uncanny ability to divide communities. He says that the polarization began in earnest in the 1930s as better communications, the radio, and travel possibilities brought local political and economic conflicts to a head. As "protest" and "folk" music went through various revivals in the 1930s and 1960s, new waves of interest from faraway music lovers and political activists brought fresh attention to the perpetual problems of the Central Appalachians.

Don Blankenship may have taken it to new heights with his media-centric Friends of America rally, in which he ma-

COAL COUNTRY CULTURE WARS

nipulated popular culture with its music videos and constant YouTube broadcasts on the Internet in a clever and sophisticated way for his own purposes.

As Mattea points out, thanks to today's short-term thinking, the vicious cycle of ducking the future won't be broken until the polarization stops. What's needed first, she says, is a truce in the culture wars of coal country.

Unfortunately, a lasting peace won't happen until there are dramatic improvements in environmental laws as well as the enforcement of those laws or the coal simply runs out. Either way, it doesn't look like there is significant change on the horizon.

9

PULLING THE TRIGGER ON DON

Bobby Ray Inman presents a curious contradiction. Dressed in slacks, a blue shirt, and a red tie with tiny stripes, he still has the erect posture of the navy admiral that he once was. Somewhat incongruously, he also has a toothy grin and Buddy Holly eyeglasses. The eightyish man maintains a folksy politeness as he greets a visitor on the thirteenth-floor office of Limestone Ventures, a small venture capital firm he runs with his middle-aged son in a modern downtown tower in Austin, Texas. His soft, friendly manner traces back to the humid East Texas piney woods where he was born and raised.

Such humble beginnings belie Inman's long career in the intelligence and military communities, where his intellect and keen tactical skills took him to the highest levels. He's been a national security adviser to five presidents; an intelligence expert; and, from 1977 to 1981, director of the ultrasecret National Security Agency, which breaks codes, eavesdrops on foreign threats, and is perhaps the federal government's largest buyer of sophisticated computer equipment. He was also the deputy director of the Central Intelligence Agency from 1981 to 1982, specializing in SIGINT, or signals intelligence, and other high-tech information-gathering methods. CIA officers I knew at the time jokingly drawled his name as "Boba Ray."

He left public service for his Texas home in 1982 and turned his laser-beam intellect to the corporate world. He was appointed a director of such marquee-name, tech-related firms as Dell Computers, software giant Oracle, Xerox, military contractor SAIC, private security firm Wackenhut, and the Carlyle Group, an investment outfit heavy with Washington insiders. For a brief time in 2011, Inman was board chairman of Xe Services, previously the Blackwater security company, made up of former navy SEALs and army Green Berets. Blackwater became notorious during the Iraq War for killing civilians. It transformed into Xe with the aim of becoming more benevolent and morphed again into a firm named Academi.

Given his deep and complex experience in national security and top-flight corporations, Inman seems an odd person to have become so intimately involved with Massey Energy, a coal firm operating in the remote hollows of Central Appalachia. Given his sterling career and the many other opportunities available to him, why did he bother to stay with Massey for so long? When I asked him about it, he responded this way: He had been a director of Fluor, a major Los Angeles military contractor and engineering firm, since the 1980s and then stayed on after the Massey spin-off. "Over time I got fascinated with the role that coal played in the production of electricity," he says. "I had been involved with Fluor since it was working all over the world and I was familiar with assessing political risk. Then, the metallurgical market [for coal] started going with China and India. It was clear that the country wasn't zooming off the nuclear [power] and I am skeptical about other alternatives." He adds that coal's importance in fast-growing Asian nations presents a strategic political challenge for the United States that is similar to ones that he had dealt with in the intelligence community. His intellectual curiosity in tracking coal's value in an increasingly global economy was just one reason why Inman stayed a director. His salary and eventual golden parachute of nearly $10 million—rather large for a director—couldn't have hurt, either.

Like his other board members, Inman was smitten with Don Blankenship's prowess at managing the tiny details of coal. The board never put together a succession plan for its top leader, which is considered necessary and routine at well-managed companies. The board tolerated Blankenship's paranoid attitudes about unions, government regulators, investment community critics, and environmental activists. Yet it was Inman who emerged as the board's powerhouse after another director—William R. Grant, who was Blankenship's biggest supporter on the board—quit following a severe stroke in 2006. By then, the board was getting tired of Blankenship's antics.

That year, Inman laid down the line, saying, "I told Don that he could either be the Republican czar of West Virginia or the chief executive officer of Massey Energy but not both." Inman believes that if Blankenship had cooled his jets, he might still be CEO of Massey today. And if he and the board were dissatisfied with Blankenship, it did not come out as they decided his compensation. Despite the flood of controversy, the board kept on grandly rewarding Blankenship with outsized pay packages worth about $30 million or more a year and goodies such as a Challenger 601 luxury jet and a house that went to Blankenship when he left the firm. In 2009, while the Upper Big Branch Mine was racking up numerous safety violations, the board paid Blankenship total compensation of $17.8 million, which was a $6.8 million increase from the previous year and double what he was paid in 2007. His deferred compensation was valued at $27.2 million.

After all, the board allowed Blankenship to build up a record of safety violations and environmental citations that were the highest in the coal industry. Inman was on the Massey board ever since it was spun off from construction giant Fluor in 1999 and while Blankenship ran amok with an intense micromanaging style that intimidated workers and fueled the wrath of the United Mine Workers of America. The board had received warning after warning demanding changes from members of the institutional investment community, such as CtW, the California Public Employees' Retirement System, and the North Carolina Retirement Systems, along with other shareholder-rights groups, but they were ineffectual. All the activists' proxy announcements, annual meeting protests, and filings with the U.S. Securities and Exchange Commission had no impact on forcing the coal firm to change its behavior. And once the recession of 2008 took hold with consequences that still last today, the heat for doing right with corporate boards automatically went on simmer. The Massey story came to a head only after the explosion.

Despite spending years as a Massey director, Inman labels the board "dysfunctional" and racked by cronyism, internal disputes, and a lack of professional governance, but his version of Massey's boardroom strife does raise significant questions as to how much influence Inman really had on the board and how hard he really tried to change things. Not being able to reel in Blankenship might be understandable if the rest of the board had been made up solely of coalfield cronies. There were enough of them, to be sure, but that wasn't the whole situation. There were high-profile directors with international reputations. Barbara Lady Judge, an American lawyer who had served as the head of the United Kingdom's Atomic Energy Authority and had been the first female commissioner at the U.S. Securities and Exchange Commission, was a director for three years. She left days after the Upper Big Branch Mine disaster in April 2010, following shareholder criticism that she had been serving on as many as six boards when four boards was considered the cutoff. Thus, she was stretched too thin to be effective at Massey, critics charged.

Another prominent director was Gordon Gee, a lawyer and academic who had been president of both Brown and West Virginia universities and is now president of The Ohio State University. He had been a Massey director from the time of the firm's spin-off from Fluor in 2000 to May 2009, when he was forced to resign under pressure from Ohio Citizen Action, an environmental group that had compiled data on Massey's record on safety and mountaintop-removal issues. The group raised hackles by complaining that Gee was in a conflict by heading Ohio State and also being a director of such a dirty and dangerous company.

If anything, Inman's predicament shows the weakness of corporate boards as they confront the need to effect change and rein in renegade executives. Not having a succession plan shouldn't prevent a board from taking the actions it needs to, although having potential leaders in the wings can help. The Massey board simply abdicated their power to Blankenship, says Professor Charles M. Elson, director of the John L. Weinberg Center for Corporate Governance at the University of Delaware. "There was no objective distance between the board and the chief executive," he says.

Indeed, Massey Energy's failed corporate governance is why the firm remained stuck in the mind-set of a bygone era despite the increasing importance of global coal markets becoming and the supposed modernization of today's corporate boards. By allowing Blankenship to be completely in charge with few checks or balances and no plan for a successor, the Massey board was in effect giving him tacit approval to act as a "robber baron, more comfortable in a boardroom in Pittsburgh in the 1890s," as he has been described by Bruce Stanley.

Inman's record as the final, lead independent director of Massey, however, is just as complicated as the rest of his adult life. It was Inman who finally pulled the trigger on Donald Blankenship and forced him to retire.

The good-old-boy way of doing business that kept Blankenship in power for so long can be traced back to Massey's history as a sleepy coal brokerage in Richmond. There, as in the coalfields, deals were made on the basis of whom you knew and how well you got along. When it was a brokerage, the firm had been run by A. T. Massey, who was described by his grandson E. Morgan as a wild, *Great Gatsby*–style rounder. After A.T.'s successors took over in the late 1940s, the company gradually started buying up reserves of coal to generate electricity or make steel. It kept its family-owned atmosphere. Managers were typically hired by handpicking individuals they knew from the coalfields who were especially knowledgeable about how to find coal, mine it cheaply, blend it, and how to install infrastructure to bring it to market. More modern managerial practices didn't seem particularly relevant in the valleys of southern West Virginia and Eastern Kentucky. If a big problem came up, it would go to Morgan, who had been a top supervisor and became president in 1972.

Back then, Massey had a reputation as a seat-of-your-pants, penny-pinching firm. It was only the catastrophic disaster at another coal company, Consol, in northern West Virginia in 1968 that brought on new federal mining laws that required coal companies to substantially upgrade safety. E. Morgan Massey says that the firm did not have the capital on its own to handle the equipment upgrades required by the new rules, so they put themselves up for sale. That, in turn, resulted in St. Joe Minerals picking up Massey in 1974, just after the Yom Kippur War in 1973 jacked up global oil prices and made coal especially desirable. Another oil-price spike after the Iranian crisis of 1979 prompted Dutch oil giant Royal Dutch Shell Group to buy half the Massey venture. The soaring price of St. Joe's coal, along with its other commodities such as gold and lead, led construction giant Fluor to buy the Massey business, giving Morgan a seat on its board.

About that time, Inman was also made a Fluor director. Besides its global engineering projects involving heavy infrastructure such as mines, highways, pipelines, and docks, Fluor was also a major defense contractor. Inman says Fluor was interested in his world perspective since they operated in many countries where politics and economics were volatile. E. Morgan was summoned to California simply because he ran a coal company but had little influence.

One consequence was that Massey was left to its own devices. It wasn't being pressured by its parent firm to update its managerial practices or double-check its safety or impact on air, rivers, and streams. Fluor, says former Massey executive and director Stanley C. Suboleski, "was a little more hands-off," although he says that Shell, which operated Massey on a 50-50 basis with Fluor, did keep some tabs on the coal firm by providing some insight into new technologies and global markets. It was during this time that Don Blankenship, an up-and-coming accountant at Rawl Sales & Processing Company, a small Mingo County, West Virginia, coal firm acquired by Massey in 1974, was hired and then taken under E. Morgan Massey's wing.

Blankenship became president of A. T. Massey in 1992. Eventually, E. Morgan Massey told me, the Fluor leadership back in Los Angeles was so impressed with Blankenship, there was talk of him eventually becoming the mother firm's CEO. Instead, he became the chief executive of Massey Energy when it was spun off in 2000. Since the new board gave Blankenship so much power, his unwillingness to negotiate just about anything would seal the company's notoriety and created conditions that led to as many as thirty-three deaths of coal miners, thousands of ruined mountaintops, and international disdain.

Another character trait of Blankenship's also affected his relationship with the board. Born without a legal father and in poverty, Blankenship was raised by his hardworking mother. He seemed in constant search for a father's advice and approval. For some time, he seemed to find them with E. Morgan Massey, who overlooked Don's rough edges, let others know of his abilities, and was backing him right up to the very moment on December 3, 2010, when the decision came for Blankenship to "retire."

In time, Blankenship would find another father figure, according to Inman. He was William R. Grant, a highly regarded executive who had been vice chairman of SmithKline Beecham and Smith Barney and later was chairman of Advanced Medical Optics and served on the boards of other firms. He had been a director of St. Joe Minerals. Later, like Inman, he had been a director of Fluor when it owned Massey. Fluor thought so highly of Grant's business acumen that it changed its board bylaws and extended the mandatory retirement age so Grant, then in his seventies, could serve as a director. According to Suboleski, another former director, Grant "was one of those old and statesmanlike people. He was staid and wise and he always had his hands around a problem. He was always reassuring people."

Like Massey, Grant was impressed with Blankenship's ability to grasp complex situations and express them in balance-sheet numbers. When Grant moved over to Massey's board when it was spun off, he quickly developed a "father-and-son" relationship with Blankenship, says James B. Crawford, a former Massey director who is now chairman of Richmond-based Carbones Inter Americanos S.A., which has coal interests in Colombia. Other members of the board included a number of Massey executives, college administrator Gordon Gee, former finance professor and Federal Reserve governor Martha R. Steger, Blankenship, and Inman, along with several others. Still, it was clear that Grant and Blankenship, who was chairman, dominated the board. The idea of a term limit or creating a framework to identify and nurture a CEO successor wasn't considered. "Bill Grant was absolutely unwilling at the outset to set some kind of succession limit," says Inman. "At the outset, if we had said six or eight or ten years, it might have been better."

Succession was just one of many sticking points on the board, which met four times a year, hopping between the five-star Jefferson Hotel in Richmond; a Marriott in Charleston, West Virginia, and in Austin; and the Greenbrier resort in southeastern West Virginia. In 2000, Massey was dealing with a $20 million fine from the U.S. Environmental Protection Agency and subsequent bad publicity for an underground mine flood in Kentucky. And while Massey seemed to be well set to prosper since Morgan and Blankenship had amassed rich troves of steam and metallurgical coal reserves, Blankenship started throwing his weight around as a highly confrontational player with regulators and a major influence in West Virginia politics.

As a director at either Massey or Fluor for more than ten years, Inman remarks that he was constantly taken aback by the poisonous atmosphere at Massey. "I haven't anywhere in the corporate world experienced the depth of the animosity dated back to the COA [Coal Operators Association] and the UMWA back in the 1980s."

"Blankenship kept sporting that TV in his office that was hit by bullets," says Inman, noting that he could never get over the fact that he was targeted in the violence. Partisan politics also were factors. The Democratic Party tended to back unions. According to Inman, "Bill Grant was a rock-ribbed Republican and he encouraged Don Blankenship. Don had grown up in that environment and he believed that on any issue you should fight. Compromise was a dirty word to him."

Another aspect of the internal culture within Massey that bothered Inman seemed to stem directly from Blankenship's personal rags-to-riches life story. One reason why Blankenship seemed so prone to confront outside regulators wasn't just because he was a political conservative who believed in limited government (although he did). "There was this feeling that if you were a private enterprise making good money, then you must be better at what you do than someone who makes less. For example, the feeling was that if you were a Massey engineer making $145,000, then you must be better than an engineer for MSHA making $80,000 a year," notes Inman. The attitude must have been especially grating for Inman, who spent thirty-two years on active duty in the navy.

211

He chose to serve as an admiral dealing with the nation's most secret and technology-heavy spy agency when he could have retired and made far more in the private sector. Blankenship, by contrast, had never served in the military or the public sector. As he would say in his many speeches to coal associations and local chambers of commerce, private enterprise was the most worthy of all endeavors. Blankenship's passionate vision of success automatically excluded the public service that Inman cherished.

Big changes, however, were finally stirring for Blankenship. Several events in 2006 brought his weaknesses to a point and eventually played a role in his ouster. One involved the emergence of two interlopers on the Massey board. New York financial celebrity Daniel S. Loeb, president of the Third Point hedge fund group, and his colleague Todd Q. Swanson got themselves on the board after investing $228.8 million in the firm, giving Third Point 5.9 percent ownership of Massey stock. Loeb declined to comment, but it seems that Inman set up the hedge fund managers as dissidents who would shake up the company. In a 2007 *Corporate Board Member* article "Hadge Funds into the Boardroom," Loeb praises Inman for helping him win seats on the board and hailed Inman's "good judgment" for putting him on the compensation and nominating committees. One man unhappy with the move was Grant. According to Inman, "Bill Grant's view was that we had to fight this guy every step of the way."

One possible reason for the Third Point play is that Inman

was subtly trying to effect change at Massey by helping tough hedge fund cowboys get seats on the board. Another view is that the wily Inman gave the hotshot New Yorkers just enough rope to hang themselves. Crawford, who joined the Massey board in 2005 and had worked for the firm when it was A. T. Massey back in the 1970s, says that the latter was the case. Loeb and Swanson were gamed by the admiral, whom Crawford describes as "a very close poker player." The two big-time financiers from New York were "renegades, young and brash, who did not have experience in coal," says Crawford. Loeb and Swanson made their play not to reform Massey but because they thought it was undervalued and they wanted to set it up for sale and make money. "Bobby Inman did a masterful job of dealing with them," Crawford says. Inman let them have their board seats, setting up their position, and later watched as they lost their seats.

About the time Third Point got two board seats, Alpha Natural Resources, a coal producer based in Abingdon, Virginia, and a Massey competitor, was pitching buying Massey. The Alpha name was new in the coalfields, but its predecessors were not. One was Pittston, another union-busting coal operator notorious for the 1972 Buffalo Creek disaster in which a badly constructed earthen dam built by Pittston burst and killed 125 residents living up and down Buffalo Creek.

As the board drama was unfolding, William R. Grant suffered a stroke that led to his death in April 2007. It left Blankenship without his most powerful ally on the board. Inman became lead director although Blankenship remained board chairman. Inman says that he wanted to take action to check the firm's direction and do a thorough assessment of how to respond to Alpha's merger offer. "I looked at this dysfunctional board and I hired Goldman Sachs to do an internal assessment about where we were and what we should do."

It was also Inman's way of trying to affect a housecleaning. "There were so many different views," he says. "You had two hostile directors and one with a stroke. That's not a functional board. You have to have the ability to develop consensus and you have to do it about 80 percent of the time. I've been on a lot of boards and I've seen this." Inman says he wanted to strengthen the audit, compensation, and governance committees of the board as it tried his upgrades. "We have to have independent board members and we need people who are financially literate," he says.

Goldman Sachs finished its assessment of whether to stay independent, accept Alpha's buyout offer, or try to buy Alpha or another company in early 2007. Blankenship was offended by Inman's review because it brought to the fore some of his quirks. Before he had a stroke, Grant was able to cover for him. "Bill Grant and Don Blankenship were almost father and son. It was that close," says Inman. It wasn't as if there were possible replacements for Blankenship, notes Suboleski, saying that one was Massey executive H. Drexel Short Jr., and another was Ben Hatfield, who was considered Blankenship's biggest potential competitor but who left Massey to work at other coal firms, including ICG and Patriot. Hatfield, according to former director Crawford, had said that any deal involving him returning to Massey Energy would have to come with a clear term limit for Blankenship. Yet Blankenship and Grant kept blunting any talk of a succession plan.

Turmoil churned behind the scenes. Outside directors Loeb and Swanson wanted Massey to sell out and were preparing a proxy battle for more outside directors—a move Blankenship fought tooth and nail. They lost their fight and resigned on June 13, 2007. To underline their point, they filed a stinging letter to Blankenship with the SEC. Loeb and Swanson said that Massey's shareholders would have been better off had the firm been sold (which would have involved dumping Blankenship). The board agreed to the sale, but "its misguided insistence on keeping you in place as CEO outweighed strategic considerations" and prevented the deal from happening. They said they repeatedly expressed concerns "about the company's business practices" and that the "company's confrontational handling of environmental and regulatory matters has simply been counterproductive." Such negatives create what Loeb and Swanson say is a "Blankenship discount" that drains away the value of company shares.

Before the resignations, Inman had his blunt chat with Blankenship about lowering his public profile, and Inman says that Blankenship did, at least for a while. But even a man such as Blankenship with a strong genius for accounting was having trouble keeping up with Massey's substantial growth during the first decade of the twenty-first century. His personal habit of micromanaging kept getting in his way. "When Massey was spun off from Fluor [in 2000]," says Inman, "it had three thousand employees and was worth $795 million. You could run every detail of it. By 2010, it was seven thousand employees and $3.5 billion. The micromanaging had become a huge issue. All of that took it out of him. When it came to coal or hedging kerosene on the market, he had no peer. Had he been able to cool the fights outside, he'd probably still be running Massey today."

With Loeb and Swanson gone and the Alpha sale blunted, Blankenship had a new lease on his Massey career. The board replaced the two hedge fund men with Lady Barbara Judge, whom Inman said could use her expertise as a top British atomic energy official to help Massey make international market choices based on its nuclear competition. Another replacement was Richard M. Gabrys, a former vice chairman of Deloitte, who had experience in the automotive and financial sectors. Their addition, Inman hoped, would improve Massey's accounting and managerial "literacy."

Gabrys and Judge ended up—along with directors Dan Moore, a West Virginia banker and car dealer, and Baxter Phillips, a top Massey executive—on the board's Safety, Environmental and Public Policy Committee. Massey was still adding to its huge list of safety and environmental violations in its deep mines and mountaintop-removal operations, and Blankenship, who wasn't as unruly as in the 2003-through-

2006 era, still relished his bad-boy rep and enjoyed tweaking activists as "greeniacs" and giving hard-edged and sarcastic speeches to business groups about regulators running amok. It wasn't clear what, if any, board-directed initiatives were in place to help the firm clean up its act.

Despite the broad experience of such directors as Inman, Judge, Gee, and Gabrys, Massey seemed completely immune to a growing movement in the corporate governance world of "socially responsible investing." Shareholder activist groups were running corporations in which large institutions had invested despite a laundry list of bad policies, such as employing child labor, polluting the air and water, putting workers in dangerous work settings, discriminating against women and minorities, and contributing to climate change. Leading the charge were such groups as Ceres, a national network that links investors with environmental interests; and the Sisters of Dominic of Caldwell; and major pension funds representing trillions of dollars, such as the California Public Employees' Retirement System, and the treasurers of Connecticut, Maryland, Vermont, Florida, and North Carolina, were reviewing their investment opportunities for how well they handle social responsibilities.

Thumbing its nose at shareholder activism, Massey Energy had become the bogeyman of the movement. Not only was its safety and environmental record seriously negative, but Blankenship's almost cartoonish propensity for confrontation made it seem that the firm was also begging to be

Exhibit A of what not to do and where not to invest. One example was the company's response to a shareholders' request that Massey consider far-reaching strategies to do something about anthropogenic climate change, noting an October 2007 study by the National Academy of Sciences had found that it was 90 percent certain that man-made greenhouse gases were to blame. Massey was asked to come up with ideas about how to mitigate the impacts of its product on climate change in a report due September 1, 2008. The company blew off the report, saying that it was "taking innovative steps to address climate change and other emissions issues." Providing no details about this policy, the firm sidestepped other issues, saying that it did not actually burn its own product and emit carbon dioxide, which had been ruled as a pollutant. The board voted down the proposal at its 2008 annual meeting. Massey continued to contribute to such lobbies as the American Coalition for Clean Coal Electricity, which had spent $1.9 million in the first part of 2008 to keep Congress from passing laws restricting carbon. The lobbies' efforts continue to be successful, especially after the 2008–09 recession, the anemic economic recovery, and the rise of the radical right-wing Tea Party movement after it stymied legislative efforts to do anything about greenhouse gases.

The ticking time bomb that was Massey's "dysfunctional" corporate governance finally went off as a secondary explosion following the one at Upper Big Branch near Montcoal, West Virginia. As Blankenship, looking downtrodden and saying little, made visits to the rescue staging areas with the flashing blue and red lights of emergency vehicles; police, state, and federal rescue teams; news media; and family and friends of the missing and dead, the board watched multiple tragedies unfold. The company had to help figure out what had happened and deal with family bereavement and the media. Suboleski says he was at his Richmond office on April 5 when Blankenship called about 3:30 P.M., or roughly half an hour after the blast. "Don said, 'We seem to have a problem. I need to know what you think,'" says Suboleski, who has a doctorate in mine engineering. At Blankenship's request, he left at once for a small civil airport where a Massey aircraft was standing by to fly him to Charleston, where a helicopter was waiting, its blades whirring. He managed to get to Upper Big Branch by 6 P.M., making a trip that usually takes more than six hours in two and a half. "I went into the command post and there were MSHA people there. They were trying to locate people but the blast happened at a shift change and that made it pretty confusing," said Suboleski.

In addition to the tragic loss of life, directors dealt with a variety of issues. Company stock was tanking quickly, its reputation was toast, and a flood of investigations and lawsuits was just about to flow in. One of Inman's first acts was to hire Public Strategies, a well-connected public relations firm just a convenient three blocks from Inman's office on Congress

Street in downtown Austin. The firm was well-suited. Former George W. Bush aides Mark McKinnon and Dan Bartlett as well as Jeff Eller, a former aide to Bill Clinton, staff it. Public Strategies' stonewalling ploy was tailor-made for Blankenship. Suboleski says that after Inman contacted Public Strategies, which merged with public relations giant Hill & Knowlton in 2011, the firm immediately sent an employee to set up shop at a Massey building south of Charleston, where he advised the board and Blankenship on how to handle the media, how to set up press conferences, and other matters. "Obviously, Don favored taking a hard line on MSHA," says Suboleski.

Massey's stock prices had been climbing back to the mid-fifty-dollars-per-share range but crashed below thirty dollars a share after the disaster. Alpha Natural Resources and other coal firms were soon to start circling Massey like buzzards sniffing carrion. In the first days after Upper Big Branch, Michael Quillen and Kevin Crutchfield, Alpha's chairman and chief executive, respectively, telephoned Blankenship to express regret for the dead miners' families and offer help with rescue efforts. On April 19, fourteen days after the blast and still in the midst of the fallout, Massey completed its nearly $1 billion purchase of Cumberland Resources and its rich and vast reserves of high-grade metallurgical coal. The deal had been hacked out over the previous year. Four days after the Cumberland deal was done, Alpha's board met to consider another pass at buying Massey. In the next several weeks, other suitors appeared, including Arch Coal, ArcelorMittal, Patriot (formerly Peabody), ICG, and several Indian mining firms.

Along with the mine catastrophe, Massey was being ripped apart internally by another specter, although the public didn't know about it. Blankenship immediately adopted an ultra defensive mode that grew quickly into paranoia, according to Inman. Blankenship feared that dissident, liberal groups such as environmental activists, along with the United Mine Workers, would somehow use the disaster and Massey's tanking stock price to take positions in the firm that would buy them board seats and force major changes.

His biggest fear, according to Inman, was that the turmoil would result in the dreaded miners' union succeeding in a massive organizing effort that Blankenship had managed to fight off for twenty years. From Blankenship's point of view, he was facing a troika of monsters out to get him. They included Richard Trumka, who had fought Massey as head of the UMWA in its watershed battle in the 1980s and was now head of the AFL–CIO. There was also Joe Main, an old foe who was now head of MSHA, and Cecil Roberts, another old enemy who was now head of the UMWA. "In Don's mind, this was all to unionize Massey."

Dealing with potential corporate governance threats became as big an issue on the Massey board as considering buyout offers and dealing with three separate probes. One was by MSHA's West Virginia mine-safety regulators, another, an independent report ordered by the then West Virginia governor Joe Manchin, and yet another by the U.S. Attorney's Office in Charleston. Alpha was pitching an all-stock transaction that would deliver a 20 percent premium at then-current prices, according to documents filed with the SEC. Blankenship balked because Massey's stock had fallen to nearly half its price from the previous winter. The dickering continued. Inman spent the summer of 2010 trying to put out fires involving corporate governance and the fears of "greeniacs" or union officials snapping up stock shares. The news media and industry analysts were having a field day. Despite the fact that Massey had a strong balance sheet, one analyst wrote—inaccurately, according to Inman—that the legal tab for Upper Big Branch would be a staggering $2 billion. "The most they ever were was $200 million," Inman says. Meanwhile, Blankenship never relented in attacking MSHA, claiming that it was involved in a "Watergate-style" cover-up of its own incompetence on matters such as ordering Massey to replace Upper Big Branch's critically important ventilation system with a design that didn't work. Blankenship, who badly wanted to keep Massey independent and maintain his job, also was trying to blunt Alpha's offer, saying that the premium it was proposing was too low to be fair to Massey shareholders, according to SEC records.

In August 2010, Massey appointed two new directors whose mission was to help Massey wade through the tidal wave of lawsuits that had been filed after the Upper Big Branch tragedy. They were Robert B. Holland III of Dallas, an oilman and financier, and Linda Welty of Greensboro, Georgia. She is a chemical engineer with experience at chemical giant Celanese along with other packaging and ink firms. Inman recalls "Don Blankenship and Linda got into quite a fight over cultural problems. They got into the issue of [safety] culture, saying it was a subjective thing and there was not enough to measure statistically. [Director] Dan Moore really got into the area of calculating risk. We thought our safety was better than what MSHA and J. Davitt McAteer, leader of the independent probe, said." James Crawford, who had been appointed head of Massey Energy's safety committee on the board in August 2009, eight months before the Upper Big Branch blast, confirms that Welty and Blankenship had at it over safety issues. "She had a background in industrial safety and was very conscious of what could have been done. In hindsight, she was right, we should have had better information on safety," Crawford says. The board, he adds, had been fed data on mine safety incidents in terms of overall averages. That made it difficult for the board to identify and track especially troubled mines such as Upper Big Branch. Incredibly, Blankenship still did not seem to realize the extent of Massey's troubles. Welty's concern about the "culture" regarding safety at Massey was too little too late. Not only was it too late to do anything about Upper Big Branch, but her influence would also be negligible.

* * *

By September, Massey's unsettled future started to take a significant turn. On September 10, 2010, Alpha sent a letter with an offer calling for a one-to-one stock change and a premium that had been increased to 26 percent. Twelve days later, Massey's board held a telephone conference to consider the Alpha offer and to start seriously considering Massey's strategic alternatives. On September 28, Massey representatives met for discussions, after which Inman and Alpha's Crutchfield started to hold one-on-one talks.

The move signaled that the admiral was steaming ahead at full speed to reach a resolution. Following the discipline of the Goldman Sachs strategic review that he had ordered in 2006, Inman bore down on a dispassionate analysis of whether to take Alpha's offer, sell part of the firm, or stay independent and even start acquiring assets again. Massey did have the option of staying independent on the strength of its exceptional coal reserves. That's what Blankenship wanted, along with director James Crawford, who represented the E. Morgan Massey way of sentimental thinking that the firm should protect its traditions, stay independent, and keep the family name. Crawford worked with Morgan at Evans Energy investments and has an office just feet away from Massey's on the first floor of the Massey Building in Richmond. Inman's ascendancy "set the stage for a process," he says, "acquiring us or taking us private or acquiring other companies." Inman says he never "wanted the leverage to go private equity."

Throughout the fall of 2010, the dickering continued. To

help sort out merger offers, Massey hired bankers Morgan Stanley and financial services firm Perella Weinberg Partners. Inman says he traveled around the country to talk with institutional investors such as Fidelity and Vanguard and asked them what they thought about a potential sale. He says that one theme constantly came up: Whatever happened, Blankenship had to go. "I visited our biggest investors and they liked Alpha, Alpha, Alpha," Inman says. As Thanksgiving approached, Blankenship was getting worn-out. He was also finally told that the constant confrontation with MSHA and other regulators had to stop. Crawford, a member of the "Morgan Massey camp of staying independent," according to Inman, said that to do so would involve firing Blankenship. "We realized that we couldn't stay independent with Don. Don had lost the support of the board," Crawford says.

On December 2, 2010, Inman flew to West Virginia and delivered the news in person to Blankenship, asking for his resignation. The following day, the board held a teleconference to decide formally to accept the resignation. Crawford recalls that he was in the Netherlands at the time on a business trip for his own coal company. Staying at the Old World–style Hotel Des Indes in The Hague, Crawford was patched into the conference about midnight, his time. The resignation was quickly accepted and the board discussed Blankenship's severance pay and other golden parachute issues. His total severance package was a stunning $86 million, according to an analysis by Cypress Associates, a New York financial house. It included $12 million in cash payments with further cash payments based on incentive and bonus programs, health care for two years, a secretary to handle his paperwork for three years, and legal aid. The board would also do very well. Inman ultimately walked away with $4.5 million.

Details of the sale were finished by the end of January 2011. Alpha agreed to buy the firm for $7.1 billion in cash and stock. Massey shareholders would get 1,025 Alpha shares plus ten dollars for each share held. The arrangement put a value on Massey shares of $69.33 a share, 21 percent more than the price of Massey's stock when markets closed January 28. Massey had $1.63 billion in debt. Yet Alpha made out extremely well in the deal. It picked up 110 mines and about 5 billion tons of coal, much of it high-heat, low-sulfur product for electrical plants and premium metallurgical coal for making coke and steel. Massey's critics said the deal was rammed through to throw up a wall around company executives to protect them from litigation.

The end of Massey Energy came on June 1, 2011, at the Marriott MeadowView atop a golf course in Kingsport, Tennessee, which, ironically, is not that far from where Blankenship moved. The sprawling conference center appeared lonely and empty as a collection of men in dark suits gathered at one of several conference rooms. The night before, Bobby Ray Inman had hosted a dinner—a kind of Last Supper—for Massey directors in a small private dining room. Hardly any shareholders were present to rubber-stamp the merger agreement. Most big institutional investors mailed in their votes. Squads of lawyers and accountants moved about the midlevel-motel ambience in a funereal silence. Within a few hours, it was over.

The board's clashes over Massey, however, flared on as the sale reached endgame. There was still fighting over the impact of the MSHA and McAteer reports that made Massey look like a shoddily run, production-obsessed, if not downright evil, monstrosity. Massey had hired its own experts to assess what went wrong at Upper Big Branch, and they supported a theory that the miners inadvertently tapped into a big reserve of natural gas. Poor safety training, ventilation, and equipment had nothing to do with the blast, it said. Inman buys the theory: "At 2:57 P.M. [the time of the blast], there was no measurable methane, then lots. That's not a slow surge." A large crevice had opened up at the end of the mine's longwall miner. Inman believes that it released natural gas.

Suboleski, who has a doctorate in mine engineering, says he was so incensed when he read the "omissions and one-sidedness" of the McAteer report that he wrote an eight-page e-mail rebutting it point by point. He sent the response to other Massey directors. Alpha Natural Resources officials were not amused. Inman wanted to release the internal Massey report to the public, and Suboleski wanted to do the same with his memo. Both could be damning to MSHA, because if natural gas was the cause, then many other mines not far from Upper Big Branch and at other locations were in big safety trouble. Alpha tried to squelch their release. "I got a letter from their lawyer asking me not to release my report. I did not," Suboleski says. But Inman did. "I got in a lot of trouble with the Alpha people," he says. Alpha, he says, just wanted to move beyond the controversies and mine coal.

Speaking for Alpha, CEO Kevin Crutchfield says his firm was peeved at Inman because "when he released that report, Massey had been Alpha's property for three days, including the report." But he admitted there wasn't much he could do about the admiral, other than have his lawyers send him a cease and desist letter.

Blankenship, meanwhile, dropped out of sight. He moved to Johnson City, Tennessee, just a few hours from his home in Mingo County, West Virginia, to be close to his son. Blankenship also shows up on occasion for depositions in some of the many lawsuits Massey Energy faces. Inman has not heard from him. "Don Blankenship's pretty bitter," Inman says. "He thinks we should have been fighting MSHA every step of the way."

What's remarkable is that Blankenship's stubborn "us-against-them" mentality dominated Massey Energy for so long. While the firm grew admirably and put together a treasure trove of high-value coal, it also earned a reputation for hardball labor and contract practices, confrontation, safety laxity, and environmental ruin. It shows what can happen when a corporate board is ill informed or inattentive or simply too busy worrying about the bottom line. In Massey

Energy's case, a tightly knit board of company insiders and distracted independent directors failed to set up a chief executive succession plan or question the data upon which they had to make life-or-death decisions. And for all their sound and fury, thumping their chests and passing out shareholders' resolutions, "green" and corporate governance activists were, in the end, ineffective. Resolution at Massey Energy came only when public outcries over the Upper Big Branch disaster finally pushed corporate directors into action.

10

ASIA'S APPETITE

Banking past tan, treeless mountains, the Air China Boeing 737 from Beijing began its final approach to Ulaanbaatar, the capital of Mongolia. My arrival into the least-populated country in the world was like entering a time machine taking me back to a provincial Soviet city from thirty years ago. On the runway tarmac sat 1960s-era Soviet-built Mi-8 helicopters along with Antonov An-2 biplanes that were being cannibalized for parts. Wafting up inside the terminal was the stench of disinfectant ammonia. Even the passport-control and customs officials were dressed in olive-brown tunics and peaked caps that seem relics of the Cold War. Just past them, hordes of scruffy taxi drivers swarmed about disembarking passengers, hawking rides to the city center that were three times the usual market fare.

The cab trip was jolting. Mongolian taxi drivers pilot their vehicles with a hell-bent-for-leather style that probably dates back nine centuries, when their horseback ancestors conquered much of the Eurasian landmass for Genghis Khan. Road rage victories are measured in the mere millimeters separating competing automobiles. Roadside scenery is a clash of culture and time. Felt-covered yurts and more recent wooden houses stand aside modern billboards pushing Mercedes, BMW, local banks, and foreign consulting firms. The acrid stench of coal smoke from the yurts and a big power plant pervades everything, burning the sinuses.

Coal is in large part responsible for the Western advertisements and the boomtown atmosphere. Besides gold and copper, Mongolia has huge amounts of both thermal and metallurgical coal prized by the fast-growing Asian countries to its south, especially China and India. How fast these countries keep growing, and what role new Mongolian coal will play, will have an immense impact over the next decade in what happens in the hollows of West Virginia and Kentucky. If demand for steel keeps up in cities like Shanghai and Mumbai, West Virginia mines will keep humming along. If Mongolia inundates the market with cheaper and easier-to-mine product, Central Appalachia's future is uncertain.

Ever since 1991 and the fall of Mongolia's Communist government beholden to Moscow, a flood of fortune seekers has swept in. They include big mining firms such as Australia's Rio Tinto and America's Peabody Coal, white-shoe financial houses such as Goldman Sachs and Deutsche Bank, consulting firms, Western law firms, chic clothiers such as Louis Vuitton, plus a slew of opportunistic entrepreneurs.

As Asia's economies roar forth, a jarring transformation has picked up speed in this remote and largely ignored nation that has about three million people and is the size of Alaska. It has been an uneven makeover. While the sidewalks are cracked and potholed, a jumble of construction cranes has erected new hotels such as the Continental and the Kempinski. Locals will shove you out of the way—Mongolians are among the world's best wrestlers. A new international school teaches the children of expatriate corporate executives. Motorists barely brake for pedestrians. Restaurants run the gamut from fatty Mongolian beef to French haute cuisine plus quaint leftovers from the years of Soviet domination, including a Ukrainian eatery serving borscht.

The treasures of Mongolia that all are seeking are located far away in more treacherous country. The prized jewel is a massive coalfield, perhaps now the world's largest, called Tavan Tolgoi, which is to the south about 400 miles and is not too far from the Chinese border about 170 miles away. "Coal is very big for us," explains Algaa Numgar, head of the nonprofit National Mining Association of Mongolia. Just five years ago, Mongolia was producing only about 7 million metric tons a year, but that shot up to 16 million metric tons in 2010 and will go to 30 million metric tons. "It will be fifty million metric tons in five years, and much of that will be from Tavan Tolgoi," says Numgar.

Indeed, some analysts regard Tavan Tolgoi as the largest new coalfield in the world—something on a par with the oil industry finding a new "super giant" trove of oil in the parlance of that industry. Of the 50 million metric tons expected from the Gobi wasteland where Tavan Tolgoi is located, about 30 million metric tons will be coking coal that will be exported to make steel. The new coalfield will automatically zoom Mongolia from virtually nowhere to third place in coking coal in the world. Leading the pack is Australia at 155 metric tons, although its loading facilities suffered badly from floods in 2010 and 2011. Number two is the United States with 51 million metric tons. Canada would be moved down to the number four spot at 27 million metric tons. And those numbers don't reflect China's demand for steam coal to make electricity, which would bring Mongolia into a handy sixth place with 23 million metric tons of global steam coal exports, nearly beating out the United States.

Mongolia has suddenly become a fascinating new possibility in the global coal industry. "The big story in the coal world is Mongolia," Robert L. Reilly, senior vice president for business development at Peabody Energy in St. Louis, told an industry conference. Peabody, which expects demand for coal to grow 90 percent over the next two decades, is one of several international coal firms, including China's Shenhua and another Russian-Mongolian consortium, that have won initial bids with Mongolia's Erdenes Tavan Tolgoi firm to develop the deposit.

Despite a lull in demand that many analysts consider temporary and related to global economic uncertainties, such as the European financial crisis, the future remains bright. China is so starved for energy, it has been opening a new coal-fired plant about every two weeks. From 2002 to 2010, according to the U.S. Department of Energy's National Energy Technology Laboratory, China grew its capacity to roughly 450 gigawatts, with a gigawatt equaling 1,000 megawatts, or enough to power 1 million Western-style homes. From 2010 to 2016, another 320 gigawatts of coal-fired electricity are planned. By contrast, the amount of coal-fired plants planned in the United States for the same period is puny—a tiny 5 gigawatts.

The incredible demand has led to unanticipated problems. In August 2010, for example, so many coal trucks were on a road between Heibei and coal-rich Inner Mongolia that they caused a traffic jam that extended sixty miles. It took ten days to sort out the mess.

Coking coal also represents a gigantic opportunity. Russia, India, Japan, and South Korea want coking coal that would be hauled from Mongolian coalfields to seaports via a new rail line that would link with the Trans-Siberian rail line. Another rail track would take coal south directly into China, bypassing truck-clogged highways and passing China's Inner Mongolia coalfields, which have become scenes of social unrest since nomadic farmers complain that coal mining is disrupting their traditional, roving lifestyles.

For American coal firms with rich metallurgical reserves such as Alpha's, Mongolia could present tough new competition. Kevin S. Crutchfield, Alpha's CEO, says that the promise of Mongolian metallurgical coalfields, along with new ones in Mozambique, "is a big strategic question for us." While Crutchfield says that it will take until at least the end of this decade for the Mongolians to develop the infrastructure to tap the country's vast coal riches, he still says it could be a game-changing event. "When they do get to that level, global trade patterns are going to change again. The Chinese will look at Mongolia as call options—where else is it going to go," he says. Australian met coal, China's import preference, would no longer be needed in as large quantities. It would have to go somewhere else, and global coal's house of cards would topple. "That means more dislocation for us or more competition in other global markets. It's something we are thinking about now," says Crutchfield.

Conversely, how Tavan Tolgoi plays out also promises to have a profound impact on how much coking coal continues to be mined in the tortured hollows of southern West Virginia and Eastern Kentucky. Depending on the speed at which Tavan Tolgoi ramps up and depending on Asia's continued demand, met coal prices could remain high. That would force more exploitation of Central Appalachian reserves and ratchet up opposition to expensive mine-safety and environmental-regulation reform. As the independent report led by J. Davitt McAteer shows, the global price of metallurgical coal drives decision making about huge investments. "The capital investment can be in the hundreds of millions of dollars for a new mine or $40 to $50 million for an existing mine. It is not uncommon for a large longwall system to be capable of producing thousands of tons per hour. Given today's pricing for metallurgical coal of $200 to $300 per ton, an operator is in a position to generate huge revenue," the report says.

As it is, coal mining in Central Appalachia has been on the decline since the late 1990s, when 300 million metric tons were produced. The amount is now about 180 million metric tons a year as coal seams become thinner, experienced miners become older and fewer, and economic and regulatory uncertainties come into play.

Should the Mongolians swamp the market with excess product, it also could turn things the other way. Lopping off more mountains in West Virginia and Kentucky would suddenly become too expensive. Adding expensive longwall miners would bog down bottom lines. That could mean more pressure by mine management to cut corners that can cost lives. It also would bring less to the local economy in terms of salaries and royalty payments.

Once more, the future of Central Appalachia, as it has been throughout modern history, will be determined by what people who live far from the region decide. In this case, it will be Asians. To get an idea of what drove the immense pressure to produce coking coal at Massey's Upper Big Branch Mine with fatal results, go visit Shanghai, a megacity with 23 million residents. The coastal metropolis is a shiny jewel of Chinese modernity, shinier even than Beijing. A century ago, it stewed as one of Asia's largest urban slums, known as the "Whore of the Orient"—a sordid fleshpot teeming with brothels, nightclubs, and opium dens. After the Communists took control in 1949, streets went dark by early evening and local residents were afraid to talk to foreigners. That changed dramatically in the 1990s after the unrest leading up to the 1989 Tiananmen Square massacre died away. Communist party bosses decided that Shanghai would transform, swanlike, into a more authentically Chinese commercial and international center, replacing anglicized Hong Kong, which would revert back to Chinese control in 1997.

The makeover was as remarkable as it was fast paced. What had been storage facilities, warehouses, and rice paddies in the once-rural Pudong District changed into the glitzy Lujiazui financial zone, with tall steel-and-glass boxes for banks, investment firms, and stock traders. Riverfront skyscrapers started soaring in the 1980s and are still stretching skyward. The Shanghai World Financial Center tops out at 1,614 feet, making it the tallest in China and the third tallest in the world. It will soon be outdone by the Shanghai Tower, which will soar 2,073 feet. Everywhere, gigantic flat-screen televisions and LED bulbs flash out a new light architecture. One example of this almost obscene aping of Western commercialism is Wujiaochang, a Shanghai square that just got a new subway stop in 2010. Four huge multilevel shopping malls surround the square. Its focal point is a passenger rail line running through the center of the square that has cladding shaped like a giant dirigible whose outer skin is covered by thousands of tiny, color-coordinated flashing lights. Even in late evening, sidewalks are thronged with shoppers all walking at different paces, bringing to mind the words of French philosopher Jean-Paul Sartre: "Hell is all the people at a Shanghai department store at the same time."

Another emblem of new China fast growth is its train system. The train to Shanghai Pudong International Airport, situated miles away in a soggy coastal marshland, zips travelers to terminals in all of seven minutes at 187 miles per hour. The train can go even faster, 268 miles per hour. It operates on a system known as magnetic levitation, in which magnetic force is used to lessen friction between rails and wheels, allowing the train to move at higher speeds. Beijing also boasts its own airport express train of a more traditional type. The destination, the Beijing Capital International Airport, is so vast that it takes ten minutes at speeds in excess of sixty miles per hour just to get from one terminal building to another. There are plenty of other examples of Chinese giganticism. The world's longest bridge over water—twenty-six miles— opened in 2011 at Jiaozhou Bay. High-speed intercity rail is popping up in remote areas, including a $3 billion project covering about 190 miles on the island of Hainan, not far from Vietnam. Just beginning is China's automobile industry, which among most industrialized countries can account for one fifth of GDP.

All these possibilities involve enormous quantities of steel and coking coal. Predictions fluctuate, but in the spring of 2011, the China Iron & Steel Associations said that steel production would be up one quarter in 2015 from where it was in 2010, rising to 750 million metric tons. Some predictions are less positive, given the sluggishness and uncertainty of the world economy, notably due to America's failure to create jobs and Europe's deficit mess over entitlement spending. Estimates are that growth in China's steel demand will drop from 4.6 percent a year to 2.6 percent a year. The overwhelming and longer-term trend, still, is upward. China's economy has grown by an average of 10 percent per year for the past thirty years.

India, too, has been gobbling up coking coal. In the first six months of 2011, steel production grew 9.3 percent, following a steady uptick in growth rates as major cities such as Mumbai and New Delhi build out. India, like China, has its own large coal industry. Indeed, Coal India has a market capitalization of $50.7 billion, second only to India's Reliance Industries, according to *The Economist* magazine.

Such miraculous growth, naturally, moves in fits and starts. The Chinese government's rush to build high-speed trains has been stymied by periodic shortages of cash, and by a major train wreck in 2011 in Zhejiang that killed forty and injured two hundred. Another dicey concept includes China's so-called ghost cities, which are being built on a massive scale. About ten planned cities a year go up. Some are the size of Reno, Nevada. Designed to alleviate overcrowding in cities such as Beijing, where as many as twenty people can share one sink and toilet, the new cities are strangely devoid of actual residents. As congestion continues in large cities, the new communities are short as many as 64 million people, although there are houses, schools, and roads ready for them.

China's pent-up growth pressure has spilled over into once-sleepy Mongolia. It is not, however, the first time developments in China, and also in neighboring Russia, have impacted the country. As always, Mongolia needs foreign money, technology, and know-how to reach and exploit its riches. It has been hampered in the past by the naïve and primitive attitudes of its leadership, a lack of law, and corruption. For instance, E. Morgan Massey considered putting his resources into Mongolia but told me he was scared away by demands for bribes.

The corruption issue has popped up before. According to professor and Mongolia expert Jack Weatherford, Mongolians faced a very similar set of circumstances in 1911, the year they became a modern country. Their response left a lot to be desired. Weatherford was speaking before a luncheon crowd of Australian and American mining experts and nattily dressed young Mongolian bankers at a conference in Ulaanbaatar of the North America–Mongolia Business Council. Canadian ambassador Greg Goldhawk introduced Weatherford, the author of *Genghis Khan and the Making of the Modern World* and a professor at Macalester College, a small but influential school in St. Paul, Minnesota. The country, he says, is on the cusp of a big surge of investment and transformation, but it had best learn from the mistakes of before, particularly avoiding corruption and shortsightedness.

In 1911, the Manchu dynasty that had dominated the land was falling apart. Mongolia's emerging leader was Bogd Khan, a monk with strange personality quirks and a penchant for high living. His flaws were easily exploited by foreign business interests that as now, had their eyes set on Mongolia's vast mineral wealth. In short order, the U.S. dollar became the de facto local currency, American Express set up an office, and German mining companies, then and now, boasted of the most advanced mining technology and were all over the rustic capital of Ulaanbaatar. Payoffs to Bogd Khan included four early-model Harley-Davidson motorcycles as a gratuity from an American company.

Russia, bordering Mongolia to its north, naturally wanted to keep its thumb on its neighbor, so the czar sent 100,000 golden rubles as a goodwill gesture. Bogd Khan spent the czar's money to outfit his government officials in the latest fashion styles. The khan also used some of his business loot to buy three hundred high-fashion dresses made in China. After they arrived, it was learned they couldn't fit stout Mongolian women. So Khan ordered his Buddhist monks to wear them. As Khan reveled in his bizarre behavior, Western firms were busy stripping the country of its resources. Between $700,000 and $1 billion in gold was taken from the country from 1910 to 1920.

What could have been a long-term tale of imperial interests corrupting and exploiting a backward nation came to an abrupt halt when Czar Nicholas II was overthrown in 1917 and then murdered with his family. That upset the regional balance of power and gave rise to Communists, who took over Mongolia in 1924, ending Bogd Khan's reign and putting a stop to the mining-concession melee as Stalin consolidated power in the new Soviet Union. Mongolia evolved as a sixteenth Soviet "republic," albeit with a veneer of independence. During the 1960s, tens of thousands of Soviet troops were garrisoned in Mongolia as tension grew between Moscow and Beijing. In 1991, the country became completely independent but is still caught within the spheres of its powerful neighbors.

The big question is what tack Mongolia will take this time. One factor, both good and bad, is handling the legacy of Soviet domination. On the plus side, a rigorous, Soviet-style primary education system gave Mongolia a literacy rate of 97.3 percent, slightly less than Argentina's but a tad better than Israel's. On the other hand, the Soviet-style command economy and backward technologies have hampered the country. Mongolia is dependent upon rail lines for shipment of goods, but much of the equipment is out-of-date Soviet locomotives and rolling stock. Trucks travel on primitive roads without many repair or fueling facilities. At the conference, T. Ochirkhuu, chairman of Ulaanbaatar Railway, explained that upgrades such as new rail lines, transfer facilities, locomotives, and rolling stock are badly needed. Replacing old Soviet diesel locomotives is critical, and the Mongolians are talking with General Electric about building an assembly plant for diesel-electric locomotives in Mongolia, where the engines must be able to withstand the 50-degrees-below-zero temperatures common during the winter.

Highways are equally lackluster, according to Sunbold Ganhold, general director of Tav Shareholding Company, a large trucking firm. Much of the newly mined coal is transported to China by truck, resulting in big problems, such as the famous ten-day-long traffic jam. Roads are inadequate and many major highways running north and south from Russia to China have no repair facilities and few gas stations. Mongolia's truck fleet is also largely obsolete. At least 40 percent of the truck engines are more than ten years old, and at least four thousand to five thousand new trucks are needed.

Mongolians are also trying to address flaws in the legal system that have encouraged the type of graft that E. Morgan Massey complained about. Bayartsetseg Jigmiddash, a young Mongolian woman who is a Harvard-trained lawyer and former advocate for the American Bar Association's Rule of Law Initiative, says that reformers are trying to get seven pieces of legislation passed to upgrade the conduct of Mongolian judges. These include having judges file statements of assets and income and report any possible conflicts of interest, says Jigmiddash, who is now advising President Elbegdorj Tsakhiagiin of Mongolia. Should the reforms be enacted, it would further momentum to bring on more transparency and rule of law. "When I started practicing law eleven years ago, it was the Dark Ages with an absolute loss of trust in the judicial system and our judges," says lawyer Bayar Budragchaa. "From 2009 to 2010, it's been a breathtaking two years. You ask, 'What the hell are these people doing?' It's because there's a generational change."

The transformation prompted by the promise of coal and other raw materials such as gold and copper is trickling down to ordinary Mongolians in other ways. One example is the stock market. In the summer of 2011, the Erdenes Tavan Tolgoi coal firm took the unusual move of giving the entire Mongolian nation shares of stock, known as TTs. The move is an obvious public relations ploy to blunt concerns about the impact of its coal mining on the country's environment, indigenous nomads, and inflation rate, but it is hard to imagine Massey Energy or Alpha Natural Resources taking the same step in Central Appalachia. Under the scheme, which is vaguely reminiscent of the Russian voucher capitalization of newly privatized state firms in the 1990s, about 10 percent of the firm's stock has been handed out. Ordinary Mongolians seem to be scratching their heads, wondering what to do about the giveaway. "They gave shares of stock to every man, woman, and child in the country," said a driver, speaking in Russian as he waited for an dignitary near a downtown building that looks like a miniature of the White House in Washington. He said he gets about eighteen dollars a month at the official exchange rate from his shares. "It's not that much, really. I use it to buy gasoline." So many other Mongolians are so confused about the stock gifts that the minuscule Mongolian Stock Exchange is ramping up an investors' education program to help the new coal shareholders understand what their shares are and if they can trade them.

While stock-share giveaways may be an oddity in the coal world, the threat of environmental degradation remains remarkably similar to what residents of other coalfields, especially those in Central Appalachia, endure. Much of the coal, notably in Tavan Tolgoi, is to be surface-mined, and the rolling landscape may make mining easier and more similar to Wyoming's grassy Powder River Basin than rugged southern West Virginia. Yet the impacts could be formidable, since, as everywhere, mining releases coal's toxic by-products in the local environment, changes the way creeks and rivers flow, and threatens wildlife.

The problems have already manifested themselves in Inner Mongolia. As China's voracious steel mills and electric power stations have been gorging on coal from Inner Mongolia, clashes have already erupted between ethnic Chinese, Chinese coal officials, and nomadic herders of Mongolian ancestry who resent having their grazing lands ripped apart by surface mines. They complain that their livestock can't eat if they have to cross new rail lines or roads. Tensions have erupted into fatal violence. One herdsman was run over and killed by a coal truck driver after protesting mining, and a forklift operator was killed during an anti-mining demonstration.

Mongolia's coal mines are in similar terrain and face problems with nomadic herders who have been living close to the land for centuries. The potential environmental and social damage is already on the radar screens of ecological activists. One is Ed Nef, a former U.S. Foreign Service officer who has been active in Mongolia for two decades in language instruction and researching the Tsaatan, otherwise known as the Reindeer People, who live in northern Mongolia near the Russian border. More recently, Nef has been working on a documentary film detailing how coal mining might impact Mongolia's environment and nomadic herders. In an eerie similarity to the propaganda wars of the Appalachians, big coal firms are already touting their respect for and management of the Mongolian environment. Peabody Energy's Web site claims it won a number of awards for its $1 million restoration of forty-four acres at the former Ereen mine near Bulgan. In what it dubs "the first coal mine restoration project," Peabody says it has returned the mine to "productive pastureland for traditional livestock grazing."

As loaded with environmental peril as China, and perhaps Mongolia, are there actually may be a silver lining in the region. It could mitigate the overwhelming damage coal causes by contributing so heavily to greenhouse gases and climate change. China is becoming a testing ground for new carbon-capture technologies. The dangers of global warming continue to build even as Barack Obama abandons serious federal regulation and as pilot plants fall by the wayside with financial problems. China does not have those woes. With its top-down Communist government, China can adopt policies quickly and without much debate. Constructing a new coal-fired plant can take a few years there, while it may take a decade or more to go through all the environmental reviews in the United States that would involve several levels of government. Also, with its gigantic export machine cranking out consumer goods worldwide, China has plenty of cash to push ahead with new coal-fired plants. Engineers believe that it is much easier to include carbon-capture technology in greenfield electricity-generating stations rather than retrofit older plants with such technology. Plus, including carbon-capture strategies in the initial design allows the Chinese the flexibility to experiment with different technologies. It is far easier and much more efficient to deploy new technology in new plants since it can be worked into a plant's original design, rather than as a tag-along retrofit.

One problem with this theory is that if political relations between Beijing and Washington flare—sticking points include currency exchange rates, intellectual property piracy, and possible Chinese aggression toward Taiwan—then sharing carbon-capture technology might come to a screeching halt. If that happens, China will have a long-term advantage over the United States and, ironically, may end up polluting far less.

Mongolia is just one of several emerging new courses of global coal. In a sense, the phenomenon invokes old-style neocolonialism, albeit with new centers of power. Mozambique in eastern African likewise has promising coal deposits, although experts say they require more cleaning and sorting. So the next shoe to drop is how quickly coal resources in places such as Mongolia can be ramped up. Algaa Numgar of the Mongolian National Mining Association says the rail lines, loading stations, and new roads should be in place relatively soon. "They will be solved in three to four years," he says. A deciding factor will be how much foreign investment the Mongolians can attract, making their efforts at legal reform all the more important. Kevin Crutchfield, the CEO of Alpha, doubts the Mongolians will have their transportation needs in place until the end of this decade. "People with big balance sheets are going to have to show up and build railroads," he says. Once they do, however, the entire global landscape for coal, especially coking coal, might be turned on its head.

The importance of coal for both electricity and steel also

could fuel geopolitical rivalries. Russia, for example, has for decades used its vast natural gas reserves as a weapon to achieve its aims in places such as Ukraine and Europe. One reason why it is so interested in investing in the Tavan Tolgoi coalfields of Mongolia is to keep its hands on a major natural resource of great interest to its neighbor and rival China. New rivalries are evolving between fast-growing India and China. The chief area of competition is for oil. Both countries use their national oil companies as spearheads for their strategic policies, and the competition is likely to spill over to other resources, such as coal, that both countries badly need to sustain their extraordinary growth.

In strict dollars-and-cents terms, the Asian-driven met coal market seems like a winner. Although coal stocks have recently taken a drubbing due to the unexpected competition from cheap natural gas, stock analysts have been betting on coal, especially metallurgical product. They believe that prices such as that of Alpha could double or triple in coming years unless the unthinkable happens and China's economy mysteriously tanks. If China's economy gets through some bumps and has a soft landing, Alpha in particular is especially well positioned, according to Dan Rice, manager of the BlackRock Energy and Resources Fund. The new metallurgical resources it picked up by buying out troubled Massey would prove a prescient move. If any one element will determine the next chapter in Central Appalachia's history, it won't be what Washington does. It will be what Beijing and New Delhi choose to do.

11

ALPHA'S ROTTEN APPLE

A wide-screen Samsung television monitor dominates a desktop wall inside a small prefabricated building at Paramount Coal's Deep Mine 41 near McClure in hilly southwest Virginia. Overlaid on a schematic blueprint of the shafts and pillars of the mine is a series of little yellow miners' helmets with an occasional locomotive icon. The helmets each depict a mine employee underground. The locomotives represent mantrips running around the shafts at the two-year-old facility that eventually will be churning out two million tons of coal a year for the next twenty years. The $350,000 locator system, based on radio-frequency-identification (RFID) technology, is designed to let managers on the surface know the exact location of everyone in the mine if there's an emergency. "We've got an exact pinpoint where everybody is and it stays on record for thirty days," says Jeff Yates, a tall, gray-headed man who is the mine's chief electrician at the facility owned by Alpha Natural Resources. Even if a catastrophe, such as an explosion, destroys parts of the system, there would still be enough data to give mine rescuers places to start looking for survivors.

Cold comfort, I thought to myself as I got ready for a tour of the mine. I was given a small rectangular plastic device, an RFID beam emitter, that I put in a pocket of my navy blue coveralls with fluorescent orange stripes on my sleeves, waist, and legs. As in previous mine visits, I was given the complete kit—a self-rescuer, a helmet with an electric light, and a battery pack—but this time they added a flashing blue strobe light in addition to the RFID beacon, plus a brief video on safety. After that we clambered onto a mantrip for the ride one mile down and 1,200 feet into the mountain. Entering the portal, I noticed that the walls and ceiling are snowy white—the result of extensive limestone rock dusting to keep down coal dust and the threat of explosions in the shaft, which is well lit by fluorescent lights. Tiny white-gray limestone particles blow like ash or radioactive fallout in the cool, steady breeze created by the mine's four powerful ventilating fans.

The ceiling is considerably higher—better than six feet—and the atmosphere less forbidding than the Massey mine I visited nearly a decade before. Within about twenty minutes we stopped and stepped out as the mantrip was parked in a side room dug out of the rock. We went through three protective plastic shields, and the floor became muddy. A miner herded us away from the shaft as a loader carrying ten tons of coal trundled from a continuous mining machine to a conveyor belt that takes the coal to the surface. The shuttle car, driven by a miner holding a box linked by cables, goes back to the mining machine, which has drumlike shearers covered with hard metal bits. A loud ripping noise is followed by a splash of water spray to keep the coal dust down. The shuttle is filled again with metallurgical coal—a process that takes about twenty-five seconds. At that time, I noticed the new antennas on the ceilings of the shafts to read radio signals such as the one beaming from my pocket transmitter.

Had Massey Energy installed a similar system at its Upper Big Branch Mine before April 5, 2010, the outcome of that disaster might have been different. It is not likely that the explosion that killed twenty-nine miners would have been prevented, but the rescue might have gone more smoothly and some lives might have been saved. That was the point Alpha Natural Resources, which bought Massey after the disaster, wanted to press home with their invitation to visit Deep Mine 41. As is typical with the coal industry, the locator systems are not put in simply due to the goodwill of company management. It is being required to do so by a federal law that was passed, once again, because of mine death and tragedy, in this case the Sago Mine disaster. The resulting MINER Act required installation of such radio locator devices at all mines by June 2011—three months too late for the twenty-nine dead at Upper Big Branch.

Alpha Natural Resources is under the spotlight for how it handles the task of remaking Massey's culture into something more modern, safer, and less intimidating. With the merger, it assumed thousands of Massey people, including many top managers, such as those who refused to cooperate with mine disaster investigators because it might violate their constitutional right against self-incrimination. Another issue is how Alpha handles closure with the families of Upper Big Branch victims. At least twenty-nine wrongful death suits were filed against Massey plus dozens of others involving safety, environmental, and contractual issues. Now Alpha has inherited them, although it has the advantage of drawing from a legal war chest that Massey officials had put together. One major deal announced in December 2011 involved the U.S. Attorney's Office in Charleston agreeing not to seek criminal prosecutions against Alpha, though prosecutors still would be free to pursue the criminal evidence that U.S. Attorney Booth Goodwin said he found against former Massey officials. In exchange, Alpha agreed to a $209 million settlement that included paying $34 million in Massey fines, investing $80 million in safety improvements that met or exceeded federal regulations at all its mines, including the former Massey properties. These would include more advanced devices to detect dangerous methane levels in mines. Another $40 million will be put into a trust to research mine safety. "The rest of the industry has no excuse but to follow this," Goodwin says, failing to note that Congress could pass laws requiring all coal firms to adopt such upgraded standards. Families of the twenty-nine dead miners and two injured miners would get a settlement of $1.5 million, according to the deal, and would still be free to pursue other restitution.

Alpha, naturally, found it in its interest to settle the lawsuits expeditiously. Still, the jury's still out on how Alpha will remake Massey's operations and rein in some of its more controversial practices, such as mountaintop removal. The more distance Alpha can get from Massey's legacy, the better, but it would be naïve to assume that Alpha will keep the entire portfolio of Massey properties forever. Top Alpha officials say that some will most likely be sold off if they don't fit strategic plans. Who eventually runs them and how are anyone's guess.

Within five months of the merger, some 6,700 people, most of them former Massey employees, had been reeducated about safety through Alpha's Running Right program. Alpha, whose total workforce is now fourteen thousand, badly needs to do this right as well. If it succeeds, it and other coal companies can start erasing years of incredibly bad publicity that Massey and Don Blankenship foisted upon the industry. By setting itself up as a twenty-first-century, well-managed coal firm, Alpha and its officials can distance themselves from the Neanderthal years of Blankenship. The tour of Deep Mine 41 was part of that public relations program. I was just one of several members of the media who have been taken on tours of the modern facility. Alpha doled out story tidbits to the regional and national media to boost its image as Massey Energy's angelic savior. A regional magazine named the firm its company of the year due to its "culture of safety." *Bloomberg BusinessWeek* likewise bought into the idea of Alpha as savior with a gushy, multipage feature.

Yet there's an inherent contradiction as Alpha officials try to make Massey fade down the memory hole. Alpha executives are simultaneously fighting a rearguard and largely unnoticed war to limit any regulatory fallout from Upper Big Branch. By contributing mightily to antiregulation and to Republican congressmen, senators, and governors and by lobbying hard, they have so far beaten back the Robert C. Byrd Mine Safety Protection Act that would introduce new regulation post–Upper Big Branch. Alpha specifically objected to parts of the bill that might make top company executives and directors criminally liable if they knew of unsafe mine practices and did nothing about it. They also have fought provisions that would give the Mine Safety and Health Administration (MSHA) the ability to subpoena witnesses if a mine incident kills three or more people. Most federal agencies already have such subpoena power. One is the Department of Agriculture, which can subpoena witnesses in far less important circumstances, such as digging up evidence of skullduggery regarding paying into a fund that promotes milk sales. The lives of coal miners are apparently not quite so important, at least according to the Washington legislative and regulatory machine. When the Mine Safety and Health Administration announced its "historic" $10.8 million in citations for Upper Big Branch, the largest ever in its history, it still paled in comparison with the $20 million in fines the U.S. Environmental Protection Agency charged Massey for an underground mine flood in 1999 in Eastern Kentucky. Asked about the discrepancy at a press conference on December 6, 2011, Solicitor of Labor M. Patricia Smith responded that according to current federal law, "It's more serious to kill a fish than it is to kill a person."

The new face of what's left of Massey Energy is that of Kevin S. Crutchfield, whose savoir faire of a world traveler is in marked contrast to Blankenship's thick-skinned hillbilly persona. The Virginia Tech mining grad has managerial experience in spots in the oil fields of the U.S. Southwest and a stint heading an Australian coal company. Crutchfield still remains true to his Central Appalachian roots, however. Alpha, formed in 2002 by the acquisitions of several traditional coal firms, is now headquartered in Bristol, Virginia, on the edge of the Central Appalachian coalfields. Lying next to the conference table at the office is an evil-looking, tan-colored, .40-caliber sniper's rifle so sturdy its barrel needs a bipod stand. "I'm into long-distance shooting," he explains, noting that he is not a hunter. The rifle can hit targets two thousand yards away or nearly two miles. "Just a hobby," he says.

Crutchfield explains that it was important to "get Running Right rolled out as fast as possible." By holding eight-hour-long classes with thirty to sixty people per class, "it took us three months but we got the majority of the legacy Massey people through it," he says. Asked how his program differs from Massey's safety training, he replies: "I really don't know how their system worked. I don't mean to be flip, but I really don't care. I think our belief was that our system was what we needed for this job. At the end of the day, to achieve an acceptable result takes two things—a system and extraordinary leadership. We have both."

Can an employee shut down an Alpha mine because of his or her own safety concerns? Crutchfield's response, "Unequivocally, yes. The issue we face in this integration is the time to strip away the fear that surrounds the issue you bring up. There is no retribution. We're after the sin and not the sinner." He can't, however, cite a specific instance in which this has happened. "That's a pretty specific question," he says. The firm does have a system to detect concerns that is based on observation cards that employees can fill out anonymously to make whatever comment they wish. "If they see something done very well, they can send it. If they see something wrong, they can submit it; they don't have to sign it," he says. All are examined, he says. "We're getting tens of thousands of cards a month. Some aren't actionable. Rome wasn't built in a day and this isn't going to happen overnight," he adds.

Indeed, not every former Massey employee has meshed

with the new system. Some have been fired. "For some, it involves a change that is not necessarily desirable from their point of view," Crutchfield says. "That's okay, we understand that. But this is the way we run the company, and this is the way we do it, and if you can't get on board with it, we'd better find another home for you. There have been discharges, yes." As to how many, he only says, "Enough."

Alpha has been under scrutiny for how many former Massey managers it has taken on. Many senior managers from Massey moved into new slots at Alpha with a few notable exceptions, including former Massey chief operating officer Chris Adkins and lead counsel Shane Harvey. Two key Massey executives who held jobs at Upper Big Branch have been given Alpha jobs. They include Chris Blanchard, the former president of subsidiary Performance Coal, which operated the mine, and Jason Whitehead. Both men went into the mine immediately after the blast although they did not have appropriate rescue training, and their presence raised the suspicions of investigators. Both took the Fifth Amendment when later questioned. In all, fifteen top Massey officials refused questions on grounds they might incriminate themselves. Asked how many were hired by Alpha, Crutchfield says, "I don't know how many. We thought a lot about that, but as you can guess, these individuals, under advice of their own counsel, suggested they take the Fifth. There's nothing criminal associated with pleading the Fifth Amendment. We obviously reserve the right in the future depending on outcomes to take additional actions." A Former Massey official, Upper Big Branch security chief Hughie Stover was taken off Alpha's payroll after he was convicted by a federal jury in Beckley in October 2011 for making false statements to investigators and was later sentenced to three years in prison and a $20,000 fine. On February 22, 2012, federal prosecutors charged Gary May, a superintendent at Upper Big Branch, with conspiracy for allegedly tipping off coworkers when regulators came to inspect the mine. About the same time, the state of West Virginia issued 253 violations against the mine, saying that foremen Ricky F. Foster and Terry W. Moore failed to clean conveyor belts or rock-dust the mine with limestone for up to four months before the blast. On March 29, 2012, May plead guilty to fraud.

Crutchfield knows that Alpha's under the spotlight on these very issues. Rumors flew that there was some kind of covert deal worked out between Alpha and Massey prior to their merger through which key Massey officials would stay on in their jobs. Crutchfield denies any such deal.

Alpha, however, has caught the eye of Congress. U.S. Representative George Miller, the ranking minority member of the House Committee on Education and the Workforce, has queried Alpha on several instances, asking about hiring former Massey officials and requesting details on its Running Right safety program. The California Democrat is also a key supporter of the Byrd mine safety bill, which was intended to introduce key reforms after Upper Big Branch. Miller has said that Alpha has "some troubling contradictions that merit a close watch." Crutchfield insists that "you can see that the remnant leadership is composed of Alpha individuals." As for Blanchard, the former president of the subsidiary that ran the Upper Big Branch Mine, Crutchfield replies, "I don't think you could call him all that senior."

Among groups closely watching Alpha is the United Mine Workers of America, which has noted that Alpha has posted signs at its facilities urging workers not to organize. While there's no indication that Alpha officials have taken the strong-arm measures that Blankenship used at Massey, the signs have stirred concern. Crutchfield admits the signs are there and says it's the company's corporate philosophy to discourage unions. "Our ongoing belief," he says, "is that people have a choice to make, but it's our voice and our recommendation that we are better served without a third party."

Like most people working in the coal industry, Crutchfield knows that the Central Appalachian fields are running on borrowed time, at least for steam coal. Although rich reserves still remain, their seams, especially those for metallurgical coal, are getting thinner and more expensive to reach. The U.S. Energy Information Administration estimates that coal production in Central Appalachia will decline 46 percent by 2015 before leveling off at about 110 million tons a year. Annual production in southern West Virginia and Eastern Kentucky will move into a "period of irreversible decline in the next several decades," reports the U.S. Geological Survey. The reason? Simple math. "Cost creep in Central Appalachia eats away at margins," says Paul Forward, a coal analyst and managing director of Stifel Nicolaus in Baltimore. So while coal prices are on the rise, the costs of mining that coal are growing even faster.

One individual who tracks coming changes closely is Rory McIlmoil of Downstream Strategies, a Morgantown, West Virginia, econometric firm and think tank. McIlmoil believes that coal's decline will be even more dramatic than some federal reports, saying it will be down 51 percent by 2020. "It's all tied to economics. It's harder to mine coal here, and our coal will be less competitive than with other basis. Powder River Basin [Wyoming] coal is expanding rapidly, and we're seeing it move farther east. Natural gas has really taken off," he says.

But not all agree with his gloomy assessment. Jerry Weisenfluh, associate director of the Kentucky Geological Survey at the University of Kentucky, says that the downward estimates don't really take into consideration metallurgical coal. While met coal won't replace thermal coal, it will mitigate the impact of its decline. Likewise, it may be too soon to write off surface mining and mountaintop removal. Weisenfluh says that federal regulators have objected to nineteen mountaintop-removal permits. If they were approved, they'd represent 120 million tons of production—a significant amount.

However the region's coal future plays out, it is clear that changes are on the horizon. Nowhere is awareness of the coming changes keener than among grassroots groups. One called Kentuckians for the Commonwealth helps publish a blog titled *Appalachia in Transition,* which provides reports, analysis, and a calendar of events related to how to move away from an economy based on coal. Its reports cover a wide range from preserving natural resources such as water supplies to enhancing farmers' markets. McIlmoil says that West Virginia needs to start preparing for the transition now.

To some extent, a lot has already occurred. The late senator Robert C. Byrd is so famous for landing federal dollars in the Mountain State that it can seem that every other road, high school, office park, and federal building is named for him, and one of Byrd's initiatives was to transform parts of northern West Virginia, where much of the coal was mined out years ago, into a technology corridor following Interstate 79 from West Virginia University at Morgantown to points south. Some of Byrd's ploys were pure pork. When the FBI needed a modern new facility to act as a depository for fingerprints, Byrd arranged to have a high-security facility built in Harrison County near Clarksburg. Called See-Jus by locals, the Criminal Justice Information Services Division employs three thousand people at its 500,000-square-foot facility, which is protected by armed guards riding its hilly perimeter in all-terrain vehicles. But with Byrd gone and budget cutting the rage in Washington, continued federal largesse is uncertain. McIlmoil says state funds should also be increased to buy new infrastructure such as roads, broadband networks, and water and sewer to help develop non-coal industries. He proposes a new mineral trust fund financed by a 1 percent increase in coal severance taxes for this purpose.

Obvious new opportunities have already taken shape. One is tourism. The New River Valley Gorge, once the horribly polluted center of West Virginia's coal and coke industry, is now a protected national park site. Several dozen companies offer white-water rafting trips on the New and Gauley rivers along with camping and rock climbing. A more recent tourism destination is the Hatfield-McCoy Trails, six hundred miles of rugged country paths for all-terrain vehicle drivers in southern West Virginia.

Despite coal's uncertain future in the region, there are surprising, if not entirely contradictory, signs that it is doing well, businesswise. Coal-industry executives may constantly complain about overregulation and its impact on employment, but the fact is that by late 2011 coal employment overall was at the highest level it had been since 1997. In West Virginia, the heart of Central Appalachia, coal accounted for 23,353 jobs, the most since 1992. But employment in the Appalachian fields is showing a curious imbalance. Demand for coal jobs is up, but the labor pool is thin. The trend will become even more acute in coming years. Many miners—up to 40 percent—will be eligible for retirement in the next several years. While companies are trying to recruit young men and some women, there aren't many miners in the middle. Alpha is trying to address this problem with a program called Generation Next, which is designed to paint coal mining as a more worthy and holistic corporate endeavor than merely being a fire boss or roof bolter or shear operator. The company is seeking younger people with college educations who can be routed both through traditionally blue-collar mine jobs and also through white-collar ones, such as engineering or human resources.

One puzzling matter in the aftermath of Upper Big Branch—and the notorious legacy of Donald Blankenship—is the dearth of real movement on mine-safety reform. The tragedy at Upper Big Branch was the deadliest underground mine disaster in the country in four decades, and any and all reform efforts have been stymied. One reason could have been that Upper Big Branch was greatly overshadowed by the subsequent massive explosion and oil spill at a British Petroleum offshore platform in the Gulf of Mexico. The huge spill, the largest ever on water, held global media attention for weeks even though one previous underground spill caused by Massey was four times as large. More telling reasons why mine-safety reforms have gone nowhere have to do with the highly partisan and polarized political atmosphere in Washington and successful lobbying by coal and other energy sectors. Crutchfield, for instance, says that legislation won't come until there is time to study and absorb what happened at Upper Big Branch.

There has been plenty already studied, however. Reports

of detailed probes have been completed by the Mine Safety and Health Administration, an independent probe ordered by the state of West Virginia, and another query by the United Mine Workers of America. They all have concluded that an overwrought corporate attitude of squeezing out profits and cutting corners at Massey Energy—plus a culture that ignored safety institutionally and penalized those who brought up such issues—were major contributing factors to the disaster. Massey's deliberate and steady way of challenging every citation hamstrung state and federal regulators and prevented a "pattern of violations" from being established that could have led to more forceful regulation. Massey conducted its own probe, but it is unique in positing that a massive leak of natural gas and not coal dust caused the blast. Although all the probes note inadequacies in regulation and with regulatory agencies, only the Massey report places much of the blame on MSHA. Among Massey's complaints is that MSHA is in a structural conflict of interest because it regulates mines and then investigates how it performed, a criticism even pro-regulation officials say has merit. In March 2012, the results of an internal probe by MSHA of its performance regarding the blast were released. The report stated that understaffed MSHA inspectors, hurt by inexperience or budget cuts, missed problems at Upper Big Branch or did not inspect problematic areas of the mine. Yet the probe states that MSHA's failings did not cause the disaster. Crutchfield says that Alpha is conducting its own review of what happened but, when pressed, he could not say when it would be completed or even if it would be made public.

The only substantive effort to bring some kind of closure to the issues brought forth at Upper Big Branch is the Robert C. Byrd Miner Safety Protection Act, first drawn up in 2010. It got nowhere in the House of Representatives, where it was drafted, and after Republicans won control of the House in November 2010, chances of it moving forward seemed nil. The bill proposes a number of regulatory changes that would prevent the kind of Massey modus operandi of thwarting efforts to cite safety infractions and enforce regulations against them. The bill would make it easier for federal regulators to take action if they can show a mine operator is chronically and repeatedly violating safety rules. Better ways of monitoring gas inside deep mines would be required. Miners who alert coal-firm management of possible safety violations would be spared retribution, as would any family member working with them in the mine.

Two key provisions in the bill have drawn the most fire from the industry. One would be to allow MSHA to subpoena testimony and documents in the case of a multiple-death incident. Incredibly, in the Upper Big Branch case, MSHA did not have the legal power to subpoena testimony. It had to ask the state of West Virginia, which does have such authority, to do it for them. As one Democratic congressional aide points out, other federal agencies have subpoena power in matters far less serious than coal mine deaths, such as allowing the federal government to subpoena witnesses and records involving a milk marketing program. Coal-industry officials are fighting hard to keep MSHA from getting any more authority. An even more controversial part of the Byrd bill would hold top-level corporate officials and directors of a coal firm or holding company criminally responsible if they learned their firm was operating in an unsafe way and they took no action. Such a regulation would go to the heart of Massey's behavior at Upper Big Branch.

For the families of Upper Big Branch victims, the issuance of another report, and the special, private explanatory sessions with the next of kin, is becoming its own routine. Massey offered $3 million to each family who lost a loved one in the disaster. The new deal between Alpha and the U.S. Attorney's Office in Charleston pays a floor settlement of $1.5 million to each of the families of the dead miners as well as to the two miners injured in the blast. Assuming that each family got both payments, they would still amount to about $130 million, or the cost of about two and a half longwall mining machines. Looking at it another way, the top executives and directors of Massey Energy were together paid a total of nearly $196 million just to go along with the sale of their firm to Alpha, according to Cypress Associates, a New York investment house who announced those numbers at a December 6, 2011 press conference in Charleston. Don Blankenship got more than $86 million. President Baxter F. Phillips Jr. got a parachute valued at $45 million, and Chris Adkins wound up with an

$11 million parachute. For his troubles guiding Massey through the days after the disaster, setting up Blankenship for his departure, and putting together the sellout to Alpha, Admiral Bobby Ray Inman got a final package worth nearly $10 million.

For the families of Upper Big Branch victims, the steady cycle of investigative reports, memorials, phone calls, and lawsuits has become routine. Some stay in touch with one another, but doing so can be difficult because the victims' relatives are spread throughout many square miles of country roads and hollows. Gary and Patty Quarles attend every information session offered to families by groups probing Upper Big Branch. The assessments of bad ventilation, inadequate rock dusting, and faulty equipment are starting to have the same ring. "There's nothing new. It's still day by day," says Gary Quarles, whose son Gary Wayne (Spanky) was killed working at the longwall face at Upper Big Branch. Their son's two children still come to visit their tidy double-wide near Naoma, but as the months drift by, the grandchildren are getting caught up with school and ball games. "We pretty much know the story by now," he says. "We pretty much have been beating MSHA to death, and the state hasn't come through. If they had true ventilation, this thing would not have happened. I forget how many ignitions they have had. At other mines run by other people, there have been something like eighteen hundred, but they didn't blow up. Without rock dusting and air, you are going to have an explosion." His wife, Patty, chimes in: "I know Don Blankenship made trillions of dollars for their company and he has millions for himself. But doesn't he have to pay for this? Isn't he going to go to jail?"

But the Quarleses, who had a wrongful death suit pending against Massey, seemed bewildered at the unexpected Alpha settlement that will pay them $1.5 million. They were among the families milling about an auditorium at an MSHA facility to hear the results of the agency's final report on Upper Big Branch. "We're supposed to be getting five hundred thousand dollars now," said Gary Quarles, wearing a black T-shirt commemorating the disaster.

Patty and me talked it over, and we'll go down to the bank when we get the check and put forty percent in for each grandchild and keep the ten percent for ourselves. If the other million dollars comes, we don't know what we'll do. We'll always have ties here, but we thought about maybe moving to South Carolina, where it's warmer. We're just people brought up with nothing, and we worked all our lives, and I spent thirty-four years underground. One million dollars may not be much money to Don Blankenship, but it's a lot of money to us.

Most of the lawsuits were settled in January 2012.

Back at Dawes, Tommy Davis arrived back at his house late one warm afternoon after work. Davis stayed on Massey's payroll for about a year after he lost his son, brother, nephew, and friends and narrowly escaped the explosion himself. After a few road trips on his Harley-Davidson and lots of thought, he decided to go back underground at another Massey mine just a few months before Massey was bought out by Alpha Natural Resources. His choice might sound illogical, but it's a story that has been repeated time and again in the West Virginia coalfields. "It's what I do and what's happened has happened," he says. As he talks, neighbors drive by his one-story house with its cluttered yard, beep their horns, and wave. It's the start of a long weekend break. Davis and his friends are thinking about a ride up to Marmet, a town east of Charleston on the Kanawha River, and hitting a bar. It's also the site of Bare Knuckles Tattoos, a market parlor patronized by Tommy and whose slogan is "Grin and bear it!" Many Upper Big Branch families have gotten tattoos commemorating their lost loved ones there. Davis has one covering his entire back. Between his shoulder blades is the name of his son with an image of a coal miner crawling in the ground. The names of his brother and his nephew flank either side along with crosses of iron spikes. At the base of his spine are crosses representing the rest of the dead with the words WAIT FOR ME IN HEAVEN and the date of the explosion.

When Tommy Davis turns philosophical, his thoughts are more personal, focused on mining, fate, and the horrible tragedy of losing his son Cory, who was only twenty years old when he was killed. "The really bad part of this is that Cory will never know what it is like to have a family," says Tommy as he stands next to the flagpole where Cory's Massey work shirt with its orange fluorescent stripes still flaps in the breeze. "He will never know what it is like to have children. He'll never know what it is like to raise them—to have them around. He died way too young. The Massey people—they are sons of bitches."

Large questions remain whether Donald Blankenship was a managerial anomaly in the twenty-first-century corporate world, how Massey Energy got away with such lax and ineffective regulatory supervision for years, and why the cycle of poverty and exploitation never seems to get broken in Central Appalachia. With bipartisan gridlock in Washington and the stubborn refusal of Republican freshmen in Congress to consider any meaningful mine-safety reforms, little will be done in the aftermath of Upper Big Branch to give the miners the protections they need. Self-seeking companies such as Alpha may enter into voluntary reform agreements with prosecutors, but these involve only one firm, not the industry as a whole. The system and regulators such as MSHA were too inept structurally and operationally to prevent Upper Big Branch, and the Alpha deal does nothing about that deficiency. Federal mine-safety regulators still won't be able to subpoena witnesses. Monetary fines simply won't be enough to change potentially deadly behavior, especially if the federal regulatory system still values a miner as less worthy of protection than a smallmouth bass or a lizard. Highly paid coal executives and their boards may still be able to dodge criminal prosecution if they know of safety issues and do nothing. And repeat offenders like Massey Energy will still be able to game the system. In short, even after the terrors of Upper Big Branch, it is still business as usual in the coalfields.

INDEX

A. T. Massey Coal Company, 37,
  39–40, 87
  divide and conquer strategy, 100
  expansion, 89–91
  secondhand equipment, 95, 97
  strip-mining, 93
  unions and, 83
Abramoff, Jack, 169
Acord, Carl, 20
Adkins, Chris, 179, 258
AEP. *See* American Electric Power
Alpha Natural Resources, 29–30,
  54–56, 76, 110
  antiregulation and, 255
  Generation Next, 264
  Massey Energy purchase, 160,
    178–79, 213, 226–27
  prices, 249
  properties, 145
  Running Right program, 160,
    254, 257
  safety and, 179–80, 253, 257

  UBB closure, 253
  unions and, 83, 161
American Bar Association Journal,
  41–42
American Coalition for Clean Coal
  Electricity, 121, 126, 218
American Electric Power (AEP),
  119–21, 129, 147
anti-coal activists, 183–84,
  187–90
Appalachia, 6, 33
  diaspora, 63–64
  drug abuse, 78–79
  economic development, 76
  education, 76–77
  folk traits, 64–68
  organized labor, 70–71
  poverty, 57–62, 271
  unemployment, 77
Appalachian coal, xv, 1, 73, 101
  decline, 260–61
  traveling to market, 111–13

INDEX

INDEX

cultural wars
  anti-coal activists, 183–84, 187–90
  Blankenship, D. in, 181–83
  Friends of America rally, 181–82,
    184–86, 197–99
  protest music, 193–96
  truce needed, 199
Cumberland Resources, 29, 220

Dam Control Act, 145–46
Davis, Cory, 5, 12, 16, 24, 31, 270
Davis, Jeff, 24
Davis, Timmy, 12, 16, 24, 64
Davis, Tommy, 5, 10–13, 64
  home of, 24
  memorial to relatives, 30–31
  philosophy, 269–71
  running for safety, 14
DeBeck, Billy, 67
deep mines, 8, 10, 96, 139
Deepwater Horizon oil spill, 56
Deliverance (Dickey), 68
Department of Energy (DOE),
  121, 123–24
Dickens, Hazel, 63
Dickey, James, 68
Disney, Walt, 67–68
DOE. *See* Department of Energy
drug abuse, 78–79
Duckworth, Brian, 185
Duke Energy Corporation,
  34, 119, 147
Duncan, Cynthia, 69

Eagle Seam coal, 8–9
Eccles Mine explosion, 70

Economist, 239
education, 76–77
electric power plants, 9, 105–6,
  129, 147
  China coal-fired plants, 234, 247
  coal and, 119–21
  demand on, 124–25
Elk Run Mine, 156–58
Eller, Jeff, 53, 220
Ellison, Terry, 22–23
Elmquist, Scott, 59
Elson, Charles M., 205
Environmental Protection Agency
  (EPA), 26, 108, 151–53, 210
Evans, Sarah Jane, 85–87
Evans, William, 84, 85

Fallows, James, 118
Farley, David, 11
Federal Coal Mine Health and
  Safety Act, 96, 167
Fifth Amendment rights, 19, 160,
  177, 179, 258
Fluor Corporation, 39–40, 101–2
  Inman at, 201–2
  Massey Energy spinoff, 216
Forward, Paul, 261
Foster, Ricky F., 259
fracking, 126–27
Frankel, Joshua, 34
Fukushima nuclear power plant
  disaster, 2, 106, 114–15
Fullen, Johnny, 43, 49–51, 55

Gabrys, Richard M., 216–17
Gallick, John, 179

INDEX

INDEX

mountaintop-removal mining, 2, 7, 26, 108. *See also* strip-mining
    aerial view, 135–38
    economic drivers, 146–47
    fast-track permitting system, 152
    land acquisition, 148
    protests, 8, 134–35, 154
    reclaiming, 48, 149
    watershed disturbance, 148–49
MSHA. *See* Mine Safety and Health Administration
Mullins, Jesse, 196–97

NAERC. *See* North American Electric Reliability Corporation
Napper, Josh, 12, 16, 24
National Mining Association, 111, 166
National Pollution Discharge Elimination System (NPDES), 151
National Resources Defense Council, 183
Nationwide 21, 152, 154–55
natural gas, 126–27, 261
Natural Resources Defense Council, 108, 125, 128
Nef, Ed, 246
New York Times, 49, 115
Night Comes to the Cumberlands (Caudill), 62, 141–43
Noonan, Peggy, 109
Norfolk Southern, 75, 78, 110, 113
North American Electric Reliability Corporation (NAERC), 124

NPDES. *See* National Pollution Discharge Elimination System
nuclear power, 106, 114–17, 124
Nugent, Ted, 181, 184–85
Numgar, Algaa, 232–33, 248

Obama, Barack, 53, 105, 151
    environmental policies, 107–9, 133, 247
    greenhouse gases and, 121–22
    weak administration, 105
Ochirkhuu, T., 243
Omar Mining Company, 89, 91
organized labor, 69–71. *See also* unions
OxyContin, 78

Paramount Coal Deep Mine 41, 250–52, 254
Peabody Energy, 29, 75, 109, 246
Performance Coal, 6, 19, 258
Persinger, Dillard, 16
petrodollars, 98
Phillips, Baxter, 103, 216, 267
Platts coal market conference, 104–8, 113–14
    close of, 127
    Massey Energy and, 129
pneumoconiosis, 26, 96. *See also* black lung disease
Popovich, Luke, 111
post-burn technologies, 118
poverty, xv, 57–62, 271
Powell, Evan, 84
Price, Joel, 16
Public Strategies, 53, 219–20